Bayesian Statistics in Actuarial Science

Huebner International Series on Risk, Insurance, and Economic Security

Other books in the series:

Cummins, J. David; Smith, Barry D.; Vance, R. Neil; VanDerhei, Jack L.: *Risk Classification in Life Insurance*
Mintel, Judith: *Insurance Rate Litigation*
Cummins, J. David: *Strategic Planning and Modeling in Property-Liability Insurance*
Lemaire, Jean: *Automobile Insurance: Actuarial Models*
Rushing, William.: *Social Functions and Economic Aspects of Health Insurance*
Cummins, J. David and Harrington, Scott E.: *Fair Rate of Return in Property-Liability Insurance*
Appel, David and Borba, Philip S.: *Workers Compensation Insurance Pricing*
Cummins, J. David and Derrig, Richard A.: *Classical Insurance Solvency Theory*
Borba, Philip S. and Appel, David: *Benefits, Costs, and Cycles in Workers Compensation*
Cummins, J. David and Derrig, Richard A.: *Financial Models of Insurance Solvency*
Williams, C. Arthur: *An International Comparison of Workers' Compensation*
Cummins, J. David and Derrig, Richard A.: *Managing the Insolvency Risk of Insurance Companies*
Dionne, Georges: *Contributions to Insurance Economics*
Dionne, Georges: *Foundations of Insurance Economics—Readings in Economics and Finance*

The objective of the series is to publish original research and advanced textbooks dealing with all major aspects of risk bearing and economic security. The emphasis is on books that will be of interest to an international audience. Interdisciplinary topics as well as those from traditional disciplines such as economics, risk and insurance, and actuarial science are within the scope of the series. The goal is to provide an outlet for imaginative approaches to problems in both the theory and practice of risk and economic security.

Bayesian Statistics in Actuarial Science

with Emphasis on Credibility

Stuart A. Klugman
Drake University

Kluwer Academic Publishers
Boston/Dordrecht/London

Distributors for North America:
Kluwer Academic Publishers
101 Philip Drive
Assinippi Park
Norwell, Massachusetts 02061 USA

Distributors for all other countries:
Kluwer Academic Publishers Group
Distribution Centre
Post Office Box 322
3300 AH Dordrecht, THE NETHERLANDS

Library of Congress Cataloging-in-Publication Data

Klugman, Stuart A., 1949-
Bayesian statistics in actuarial science : with emphasis on credibility / Stuart A. Klugman.
p. cm. — (Huebner international series on risk, insurance, and economic security)
Includes bibliographical references and index.
ISBN 0-7923-9212-4 (acid-free paper)
1. Insurance—Statistical methods. 2. Bayesian statistical decision theory. I. Title. II. Series.
HG8781.K58 1991
368'.01—dc20 91-25856
CIP

Printed on acid-free paper.

Printed in the United States of America

For Marie

TABLE OF CONTENTS

PREFACE

The debate between the proponents of "classical" and "Bayesian" statistical methods continues unabated. It is not the purpose of the text to resolve those issues but rather to demonstrate that within the realm of actuarial science there are a number of problems that are particularly suited for Bayesian analysis. This has been apparent to actuaries for a long time, but the lack of adequate computing power and appropriate algorithms had led to the use of various approximations.

The two greatest advantages to the actuary of the Bayesian approach are that the method is independent of the model and that interval estimates are as easy to obtain as point estimates. The former attribute means that once one learns how to analyze one problem, the solution to similar, but more complex, problems will be no more difficult. The second one takes on added significance as the actuary of today is expected to provide evidence concerning the quality of any estimates.

While the examples are all actuarial in nature, the methods discussed are applicable to any structured estimation problem. In particular, statisticians will recognize that the basic credibility problem has the same setting as the random effects model from analysis of variance.

As the technological problem is the most difficult, this monograph spends a great deal of time on computational issues. The algorithms are presented in the text and computer programs are listed in the Appendix. In addition, each model is illustrated with an example that demonstrates that these algorithms can be applied to realistic data sets. The data sets are also included in the Appendix so that the interested reader can duplicate the results presented here as well as try other models.

This work had its origins when I spent the Fall of 1984 at the National Council on Compensation Insurance in New York City. I learned a great deal about credibility while working on a rate making project there. This project was headed by Gary Venter, from whom I learned a great deal about credibility and with whom I had numerous conversations over the following years. Gary also was kind enough to supply the data that appears in Data Sets 2–4. Many other useful ideas came from conversations with Glenn Meyers, particularly while we were faculty members at The University of Iowa. I, of course, owe an enormous debt to the many actuaries who from the beginning of the century have advanced the theory of credibility and to the many statisticians who have moved Bayesian analysis from theory to practice.

Finally, financial support for this project was given by the S. S. Huebner Foundation. This provided the time and the incentive for completing this project.

Bayesian Statistics in Actuarial Science

1. Introduction

In the minds of most statisticians there are (at least) two mutually exclusive approaches to data analysis. The "classical" or "frequentist" theory consisting of confidence intervals and hypothesis tests is certainly the most widely used and takes up the vast majority, if not all, of the typical statistics text. On the other hand, "Bayesian" statistics, a mode of inference based on Bayes' Theorem, has attracted a small group of passionate supporters. The debate continues with papers such as "Why Isn't Everyone a Bayesian?" (Efron, 1986) drawing numerous comments and letters.

The purpose of the monograph is not to settle this issue or even to attempt to make a contribution toward its resolution. Rather, it is to demonstrate that there is a well-deserved place for Bayesian statistics in the analyst's arsenal. This is accomplished by showing that two important actuarial problems can be successfully solved by the Bayesian approach. The first is the general problem of model-based prediction and normally involves three steps. Begin by collecting data on the past behavior of an insurance system. This may, for example, be the ages at death of the holders of a certain type of life insurance policy or the indemnity amounts paid to the members covered by a group insurance policy. The second step is to build a model describing the random process that yielded these observations. Examples of such models abound in the actuarial literature (e.g., Batten (1978), Hogg and Klugman (1984), Hossack, Pollard, and Zehnwirth (1983), and London (1988)). The third step is to make predictions of future outcomes based on the model. Of growing importance to the actuary is not just the expected future outcome, but the variability of the prediction. In measuring variability it is essential that as many sources as possible be taken into account. The two most obvious are the natural process variability that is inherent in the model and the variability because the model is estimated from data. It will be seen that the Bayesian approach provides an ideal mechanism for combining these two sources of variability.

The second problem comes under the general name of credibility. Since the first part of the twentieth century it has been recognized that when setting insurance premiums the best result is obtained when the pure premium is somewhere between the actual experience of the insured and the overall average for all insureds. This can be justified in several ways. Foremost is that an individual insured or group of insureds will rarely provide enough information to accurately estimate their potential for future claims. As a result there will be wide spreads in the recorded experience from year to year. By shrinking to the overall average these fluctuations will be tempered. Another advantage is that insureds with exceedingly unfavorable experience in a given year will not be penalized to the full extent indicated by this experience. (Of course this must be balanced by giving a less than full reward to those with good experience.) This creates less hostility among the policyholders. At the other extreme it is clearly incorrect to set a pure premium of zero for an insured who has no claims during the experience period. The difficult problem is the determination of the relative emphasis to place on the specific and overall experiences. Again, it will be seen that a Bayesian approach is the best way to resolve this problem.

The first step in solving these problems is to provide the specifics of the Bayesian approach to modeling and data analysis. This is done in Chapter 2 where it is seen that the ideas are deceptively simple while the implementation is decidedly complex. Most of the complexity centers on the computations needed to complete the analysis. For the most part these turn out to be numerical integration problems. Solutions to the computation issues are provided in Chapter 3.

The problem of model-based prediction is taken up in Chapter 4. It begins with a simple model and an appropriately simple illustration. This is followed by a more complex example. This Chapter also introduces the Kalman filter, a powerful model for describing an environment in which parameters change over time. This allows the introduction of a third kind of uncertainty, that tomorrow's model might be different from yesterday's.

The credibility problem is attacked in Chapters 5 through 9. The details of the problem are set out in Chapter 5 along with various issues that the analyst must resolve. The advantages of a Bayesian analysis are detailed as part of this discussion. The next step is to simplify matters by placing a number of assumptions on the claims process. While they may appear to be unrealistic in the insurance setting, using them leads to results that match those currently being used. These assumptions are stated and defended in Chapter 6. In Chapter 7 the basic Bayesian model is introduced. It incorporates the assumptions set out in Chapter 6 and adds the additional assumption of linearity. Many of the commonly used models fit this framework. The formulas for doing a complete credibility analysis are derived and the formulas for a number of specific examples given. In Chapter 8 several data sets are analyzed to demonstrate both the power and feasibility of the Bayesian approach. Finally, in Chapter 9 the assumptions introduced in Chapter 6 will be relaxed to accommodate more models. In this chapter the power of the Bayesian approach will be fully realized. The monograph will then close with an Appendix in which programs for performing the various computations and a description of the algorithms are given.

Throughout the monograph a number of examples will be used to illustrate the various methods. Their main purpose is to demonstrate that the proposed methods can be implemented. To avoid introducing a new example for each model it will often be the case that a data set is analyzed with a model that is clearly inappropriate. One benefit is that the model selection techniques introduced in Chapter 2 should be able to point this out. Nevertheless, it should be recognized that model selection may well be more important than the method used to analyze the model. Selection of a good model depends on a thorough knowledge of the process that produced the losses.

All the computations were done on an MS-DOS based personal computer (80286) with the numerical co-processor added. All the programs were written in the GAUSS programming language and run under that system. This language is similar to the SAS Matrix language.[1] More details are given in the Appendix.

[1]MS-DOS is a trademark of Microsoft, Inc.; GAUSS is a trademark of Aptech Systems; Inc., SAS is a trademark of SAS, Inc.

2. BAYESIAN STATISTICAL ANALYSIS

A. THE BAYESIAN PARADIGM

As discussed in Chapter 1, it is not the intention of this monograph to provide a convincing philosophical justification for the Bayesian approach. Excellent discussions of these matters can be found, for example, in Berger (1985, Chapters 1 and 4) and Lindley (1983). In the latter paper the Bayesian paradigm is described in its simplest form. Of interest is a quantity θ whose value is unknown. What is known is a probability distribution $\pi(\theta)$ that expresses our current relative opinion as to the likelihood that various possible values of θ are the true value. For additional discussion of the merits of expressing uncertainty by probability see Lindley (1982 and 1987). This is called the prior distribution as it represents the state of our knowledge prior to conducting the experiment.

The second item is the probability distribution $f(x \mid \theta)$. It describes the relative likelihood of various values of x being obtained when the experiment is conducted, given that θ happens to be the true parameter value. This is called the model distribution and is the one element that is common to both Bayesian and classical analyses. It should be noted that it is possible for both x and θ to be vectors, the former being the several items from the sample and the latter a set of unknown parameters.

The next step is to use Bayes Theorem to compute

$$\pi^*(\theta \mid x) = \frac{f(x \mid \theta)\pi(\theta)}{\int f(x \mid \theta)\pi(\theta)d\theta} \tag{2.1}$$

the posterior distribution[2] of θ. It represents our revised opinion about θ having seen the results of the current experiment. This density contains all of our current knowledge about the unknown parameter, just as the prior density used to.

The final step is to use the posterior density to draw whatever conclusions are appropriate for the problem at hand. We will concentrate on four specific items. The first is a point estimate of one of the parameters in the vector θ. Writing it in vector form, we have $\boldsymbol{\theta} = (\theta_1,\ldots,\theta_k)'$ and the posterior distribution of θ_i is

$$\pi^*(\theta_i \mid x) = \int \pi^*(\theta_1,\ldots,\theta_k \mid x)d\theta_1\cdots d\theta_{i-1}d\theta_{i+1}\cdots d\theta_k. \tag{2.2}$$

A viable point estimate of θ_i is then some measure of the center of this posterior density. The two most common are the mean and the mode. Justification for these choices can be obtained by choosing appropriate loss functions and applying Bayesian decision theory, but it is sufficient to recall that the posterior is the sum of our knowledge and the mean and mode are useful summary values. The second item is the Bayesian version of the confidence interval. A $100(1-\alpha)\%$ credibility (not to be confused with actuarial credibility as discussed in Chapters 5 through 9) interval for θ_i is any pair (l,u) that satisfies

$$Pr(l < \theta_i < u) = \int_l^u \pi^*(\theta_i \mid x)d\theta_i = 1 - \alpha.^3 \tag{2.3}$$

Perhaps the most desirable choice for the end points is that pair for which $u-l$ is the smallest. For a unimodal density the solution is the pair that satisfies both (2.3) and

$$\pi^*(\theta_i = l \mid x) = \pi^*(\theta_i = u \mid x). \tag{2.4}$$

---

[2]Throughout this monograph all densities will be written as if the random variable is absolutely continuous. For discrete variables the integrals should be replaced by the appropriate summation.

[3]The often used convention of using upper case symbols for random variables and lower case symbols for observed values will not be used here. Only one version of the symbol will appear and its role should always be obvious from the context. As well, the words estimate and estimator will be used interchangeably. The Bayesian analyst must be comfortable with the concept that both parameters and observations are random variables at times.

The Bayesian Central Limit Theorem (Berger, 1985, p. 224) indicates that under suitable conditions the posterior density can be approximated by the normal distribution and so the credibility interval is approximately

$$E(\theta_i \mid x) \pm z_{1-\alpha/2}\sqrt{Var(\theta_i \mid x)}. \tag{2.5}$$

As with the usual Central Limit Theorem, the approximation improves as the number of observations increases.

Of greater interest is the value of a future observation. Suppose the density of this observation is $g(y \mid \theta)$. Note that the model for this density need not match the model that produced the observations x, but it must depend on the same parameter θ. The predictive density is then

$$f^*(y \mid x) = \int g(y \mid \theta)\pi^*(\theta \mid x)d\theta \tag{2.6}$$

and it represents all of our knowledge about a future observation. Both point and interval estimates for this value can be constructed as was done for the parameter itself.

There are three major problems in implementing the Bayesian paradigm. One is the selection of the prior distribution. This is the most often criticized facet of Bayesian analysis with the objection being made that the personal nature of this distribution removes the aura of objectivity that is expected to surround scientific explorations. The second is the selection of the model. The problems are not much different from those faced by all analysts. These two issues will be discussed in detail later in this Chapter. Once the two elements are in place the analysis then consists of evaluating the appropriate expressions as given above. While the formulas are simple in appearance, the required integrals are often extremely difficult to do. Being able to perform high dimension numerical integration is essential for the Bayesian analyst. Several techniques for doing this are presented in the next Chapter.

B. AN EXAMPLE

At this point it is appropriate to introduce an example. Suppose we are observing the number of claims from one randomly selected driver from the collection of all insured drivers. Assume that the number of claims in one year has a Poisson distribution with parameter θ and that the numbers of claims in different years are independent. If we observe t years of claims from one insured the model is

$$f(\boldsymbol{x}|\theta) = e^{-t\theta}\theta^{\Sigma x_i} \Big/ \Pi x_i!. \tag{2.7}$$

Further suppose that our prior distribution is the gamma distribution with parameters α and β. That is

$$\pi(\theta) = \frac{\theta^{\alpha-1}e^{-\theta/\beta}}{\beta^{\alpha}\Gamma(\alpha)}. \tag{2.8}$$

One way to interpret this distribution is that it represents the way in which the values of θ are spread among all the drivers in the population of potential insureds. Before any data have been obtained on this particular driver all we know is that he is a member of this population. Discovering the values of α and β may not be easy, but this problem will be ignored for now.

The numerator of the posterior distribution is

$$\pi^*(\theta|\boldsymbol{x}) \propto \frac{e^{-t\theta}\theta^{\Sigma x_i}}{\Pi x_i!}\frac{\theta^{\alpha-1}e^{-\theta/\beta}}{\beta^{\alpha}\Gamma(\alpha)} \propto \theta^{\alpha+\Sigma x_i-1}e^{-(t+1/\beta)\theta}. \tag{2.9}$$

The denominator of (2.1) does not depend on θ and so must be the constant that will make (2.9) a density function, that is, integrate to one. From inspection we have a gamma distribution with parameters $\alpha + \Sigma x_i$ and $\beta/(1+t\beta)$. The Bayes estimate of θ is the posterior mean, $(\alpha + \Sigma x_i)\beta/(1+t\beta)$. This can be rewritten as

$$\frac{t\beta}{1+t\beta}\bar{x} + \frac{1}{1+t\beta}\alpha\beta,$$

a weighted average of the sample mean and the prior mean. It is noteworthy that as the amount of data (t) increases so does the weight placed on the sample mean. In the insurance context, the more years of experience available on this particular driver, the more reliable it is for setting future premiums for that driver.

We can also evaluate the precision of this estimate using the posterior variance

$$Var(\theta \mid \boldsymbol{x}) = \frac{(\alpha + t\overline{x})\beta^2}{(1 + \text{t}\beta)^2}. \tag{2.10}$$

Now assume that a future observation y will be obtained from the same Poisson distribution. The predictive density is

$$f^*(y \mid \boldsymbol{x}) = \int_0^\infty \frac{e^{-\theta}\theta^y}{y!} \frac{\theta^{\alpha+\Sigma x_i - 1} e^{-(t+1/\beta)\theta} (t + 1/\beta)^{\alpha + \Sigma x_i}}{\Gamma(\alpha + \Sigma x_i)} d\theta$$

$$= \int_0^\infty \frac{\theta^{\alpha+y+\Sigma x_i - 1} e^{-(t+1+1/\beta)\theta} (t + 1/\beta)^{\alpha+\Sigma x_i}}{y!\Gamma(\alpha + \Sigma x_i)} d\theta$$

$$= \frac{(t + 1/\beta)^{\alpha+\Sigma x_i} \Gamma(\alpha + y + \Sigma x_i)}{(t + 1 + 1/\beta)^{\alpha+y+\Sigma x_i} y!\Gamma(\alpha + \Sigma x_i)}$$

$$= \frac{\Gamma(\alpha + y + \Sigma x_i)}{y!\Gamma(\alpha + \Sigma x_i)} \left[\frac{t + 1/\beta}{t + 1 + 1/\beta}\right]^{\alpha+\Sigma x_i} \left[\frac{1}{t + 1 + 1/\beta}\right]^y \tag{2.11}$$

which is the negative binomial distribution. We could get the moments of y either directly from this negative binomial distribution or from

$$E(y \mid \boldsymbol{x}) = E[E(y \mid \theta, \boldsymbol{x})] = E(\theta \mid \boldsymbol{x}) = \frac{(\alpha + \Sigma x_i)\beta}{1 + t\beta} \tag{2.12}$$

$$Var(y \mid \boldsymbol{x}) = E[Var(y \mid \theta, \boldsymbol{x})] + Var[E(y \mid \theta, \boldsymbol{x})$$

$$= E(\theta \mid \boldsymbol{x}) + Var(\theta \mid \boldsymbol{x}) = \frac{(\alpha + \Sigma x_i)\beta}{1 + t\beta} + \frac{(\alpha + \Sigma x_i)\beta^2}{(1 + t\beta)^2}. \tag{2.13}$$

To make the example explicit consider a prior distribution with $\alpha = .3$ and $\beta = 1$. Further suppose that in the past ten years the applicant had 0, 1, 0, 2, 0, 0, 0, 1, 0, and 0 claims. The predictive distribution has mean

$$(.3 + 4)(1)/(1 + 10) = .39091$$

and variance

$$.39091 + (.3 + 4)(1)^2/(1 + 10)^2 = .42645$$

for a standard deviation of .65303. These are exact; any approximation is due to the prior distribution or the choice of the Poisson model.

Contrast this with the classical analysis for this problem. The maximum likelihood estimator of θ is the sample mean, $\overline{x}$. In this situation it is not difficult to find the distribution of the estimator, namely $n\overline{x}$ has the Poisson distribution with parameter $n\theta$. Arguing as is done in the usual construction of prediction intervals in the regression setting we have, for the next observation, y, that its estimator $\overline{x}$ will vary from it by

$$E[(y - \overline{x})^2 \mid \theta] = E[(y - \theta + \theta - \overline{x})^2 \mid \theta]$$

$$= Var(y \mid \theta) + Var(\overline{x} \mid \theta) = \theta + \theta/t. \tag{2.14}$$

At this point it is customary to approximate the parameter by its estimate and announce the variance as $\overline{x}(1 + 1/t)$. We have little information about the accuracy of this approximation in that any study would again produce an answer that depends on θ.

To close, note what happens in (2.12) and (2.13) if we let $\beta \to \infty$ with $\alpha\beta = \mu$ held constant. The Bayes predictive density will have a mean of $\overline{x}$ and a variance of $\overline{x}(1 + 1/t)$. This particular prior distribution has an infinite variance. It is one way of representing a near total lack of prior knowledge, and so in some sense ought to produce a result similar to the classical offering. However, here we obtain this result without having to resort to an unsatisfying approximation. This type of prior distribution will be explored in greater detail in the next section.

C. PRIOR DISTRIBUTIONS

The example in the previous Section closed with an illustration of what is often called a vague or noninformative prior. This term refers to any prior distribution that attempts to display a near total absence of prior knowledge. A more constructive view is to state that all relevant prior information has been incorporated into the description of the model. The hierarchical models introduced in Chapter 6 and discussed in detail in Chapter 7 are especially useful when the model must contain several pieces of information.

Unfortunately there is no agreement as to what it means for a prior to be noninformative or how to find the best one in a given situation. Bernardo (1979) shows just how difficult it can be to devise an automatic method for the construction of noninformative priors. The discussions at the end of that paper also indicate how much heated debate this topic can generate.

To continue the illustration, let us see what happens to the gamma density as $\beta \rightarrow \infty$ with $\alpha = \mu/\beta$.

$$\pi(\theta \mid \alpha,\beta) = \frac{\theta^{\alpha-1}e^{-\theta/\beta}}{\beta^{\alpha}\Gamma(\alpha)} = \frac{\theta^{\mu/\beta-1}e^{-\theta/\beta}}{\beta^{\mu/\beta}\Gamma(\mu/\beta)} \rightarrow \frac{\theta^{-1}(1)}{(1)(\infty)} = 0 \tag{2.15}$$

an unsatisfactory choice for a density. There are, however, two prior distributions that do fit the notion of being noninformative. One is the "density"

$$\pi(\theta) = 1, \qquad \theta > 0. \tag{2.16}$$

Of course, it is not a density at all since it does not integrate to one, in fact, the integral does not exist. However, it does have an appealing appearance, namely uniformity. It appears to express the prior opinion that one value of θ is just as likely to be correct as any other. Employing this prior in the previous analysis yields

$$\pi^*(\theta \mid \boldsymbol{x}) \propto \frac{e^{-t\theta}\theta^{\Sigma x_i}}{\Pi x_i!} \propto \theta^{\Sigma x_i}e^{-t\theta}. \tag{2.17}$$

From inspection we have a gamma distribution with parameters $\Sigma x_i + 1$ and $1/t$. The Bayes estimate of θ is $\overline{x} + t^{-1}$. The posterior variance is

$$Var(\theta \mid \boldsymbol{x}) = (t\overline{x} + 1)/t^2. \tag{2.18}$$

The predictive density is

$$f^*(y \mid \boldsymbol{x}) = \frac{\Gamma(1 + y + \Sigma x_i)}{y!\Gamma(1 + \Sigma x_i)}\left[\frac{t}{t+1}\right]^{1+\Sigma x_i}\left[\frac{1}{t+1}\right]^{y} \tag{2.19}$$

which is again a negative binomial distribution. All the results from the previous section can be duplicated by setting $\alpha = 1$ and $\beta = \infty$. The most interesting observation is that even though the prior density is not really a density at all, useful results still obtain from applying Bayes rule. This is often, although not always, the case when such noninformative priors are used.

For this particular example the uniform prior is not very satisfying. Its mean is infinite, which is certainly not anyone's idea of where the concentration of prior opinion should be. Another problem is that the Bayes estimate is always larger than $\overline{x}$, since it is a compromise between $\overline{x}$ and the infinite prior mean. A popular alternative noninformative prior for a one-dimensional parameter is the square root of the Fisher information[4]. For the Poisson distribution it is

$$\pi(\theta) = \theta^{-1}. \tag{2.20}$$

This is equivalent to using $\alpha = 0$ and $\beta = \infty$. The posterior density of θ is now gamma with parameters $t\overline{x}$ and $1/t$ for a posterior mean of $\overline{x}$. The predictive distribution is now negative binomial with parameters $t\overline{x}$ and $t/(1+t)$. The mean and variance are $\overline{x}$ and $\overline{x}(1+1/t)$ the same result as in the previous section. This seems to reinforce (2.20) as a good choice for a noninformative prior.

Although θ is not a scale parameter for the Poisson distribution, there are some compelling arguments that (2.20) is always the appropriate noninformative prior for scale parameters. When estimating a scale parameter (such as the standard deviation) any reasonable noninformative prior should have the following property: Consider two individuals who profess to have the same prior opinion about a process. If one individual observes X from scale parameter β and a second individual observes Y from scale parameter $\gamma = 2\beta$ then (1) if $Y = X$ the first individual's posterior concerning β should be identical to the second individual's posterior concerning γ and (2) the first person's prior probability on a given set of values should be equal to the second person's prior probability for the equivalent set where all the values have been doubled. The second requirement just says that the two individuals have the same prior opinion about their respective problems. The first requirement reflects the

[4]It is often called the Jeffreys prior (Jeffreys, 1961). The Fisher information is the negative expected value of the second partial derivative with respect to the parameter of the natural logarithm of the density function. An excellent discussion of this prior and noninformative priors can be found in Box and Tiao (1973, pp. 41-60).

noninformative nature of the prior. If we really do not have any information, being told, for example, that we are recording observations in cents instead of dollars should not change the answer. It is easy to show (see Berger, 1985, pp. 83-87 for a more detailed discussion) this implies the prior density must satisfy the relation $\pi(\beta/\gamma) = \pi(1)\pi(\beta)/\pi(\gamma)$ for all β and γ. Such a prior is called relatively scale invariant. A class of functions that satisfies this relation is $\pi(\beta) = \beta^p$ where p is any real number. One way to narrow the choice of p is to further restrict the definition of scale invariance. Suppose in restriction (1) the requirement is changed to demand equal priors instead of equal posteriors. In that case the only member of the family that still qualifies is $\pi(\beta) = 1/\beta$. This prior has another desirable property that makes it unique in this class. Select an arbitrary value, say v. Then compute $\int_0^v \beta^p d\beta$ and $\int_v^\infty \beta^p d\beta$. For all $p \neq -1$ one of these integrals is finite while the other is infinite, indicating that a preponderance of the prior probability is either being placed on very large values ($p > -1$) or on very small values ($p < -1$). Only with $p = -1$ will both integrals be infinite, thus indicating a true indifference between large and small values.

Additional problems arise when there are multiple parameters. If the prior opinions are independent, then the individual noninformative priors can just be multiplied. In more complex settings either more clever approaches are needed (see Chapter 7 for some examples), or the all-purpose prior

$$\pi(\theta_1,\ldots,\theta_k) = 1 \tag{2.21}$$

can be used. It is important to note that if the situation is such that the choice of noninformative prior has a profound influence on the eventual results, then the data must be contributing very little. At this point the analyst should either get more data, get a better model, spend the time to formulate an informative prior, or give up and recognize that there is just not enough information available to allow a reasonable conclusion to be drawn.

D. MODEL SELECTION AND EVALUATION

When several models are available for the same situation it becomes necessary to have a mechanism for determining the one that is most appropriate. This is especially true for nested models. In the Poisson example an alternative model is the negative binomial. This is not really a case of nesting as the Poisson is a limiting and not a special case of the negative binomial. Even if only one model is being considered, its validity should still be verified. In this Section two techniques will be presented; the first is a graphical procedure that is similar to standard regression diagnostics and the second is similar to a likelihood ratio test.

1. Graphical Analysis

The standard technique for evaluating the assumptions used in a regression analysis is to plot the residuals against the fitted values. For a Bayesian analysis the same concept applies. Box (1980 and 1983) also suggests looking at the marginal distribution as found from the estimated prior distribution and then seeing if the observations appear to be a random sample from that distribution.

One way to do this is to look at the model distribution conditioned on the estimates of the various parameters, that is

$$f(x \mid \hat{\theta}). \tag{2.22}$$

In the example with the noninformative prior $\pi(\theta) = 1/\theta$ we have $\hat{\theta} = \overline{x} = .4$ and the estimated model density is Poisson(.4). We then must find a graphical display that will indicate whether or not the ten observations might have come from this distribution. The most convenient way to do this is with a histogram or a probability plot. To make a probability plot the n data points are arranged in increasing order and n equally spaced percentiles from the model distribution are also obtained. The usual choice for equal spacing would be the $1/(n+1)$, $2/(n+1)$, ..., $n/(n+1)$ percentiles. The n pairs of points (one from the data and one from the list of percentiles) that result are then plotted. If the model is a good one the points should lie near the forty-five degree line emanating from the origin.

With just ten data points and a discrete distribution, this technique will not be of much value for the example. The ordered observations are 0, 0, 0, 0, 0, 0, 0, 1, 1 and 2 while the percentiles are 0, 0, 0, 0, 0, 0, 0, 1, 1 and 1 (the probability of getting 0 is .670 so the first 7 percentiles, through $7/11 = .636$, are 0, and the probability of getting 1 or less is .938 which covers the last three, the final one being $10/11 = .909$). Of course many other discrete distributions will also show this level of agreement.

Some more useful examples of this procedure will be presented in later Chapters.

2. A Selection Criterion

There are many ways to test the aptness of a model. The one presented here is the Schwartz Bayesian criterion (Schwartz, 1978). The basic idea is to look at the joint density (model times prior) and then insert estimates of the parameters and the data. Next take the natural logarithm and subtract p times the natural logarithm of $n/2\pi$ where p is the number of parameters in the model and n is the sample size. When two models are under consideration, the one with the largest value is to be considered more likely to represent the process that yielded the observations. The subtraction imposes a penalty on models that have a large number of parameters.

There is one problem with implementing this approach. Suppose one model is a special case of another one. When looking at the model's contribution to the joint density, there is no doubt that the more complex model will produce a larger value. However, the method also requires the addition of the natural logarithm of the prior density. If radically different priors are used for the two situations, they will affect the result. This makes good sense if we are using informative priors, but not when noninformative priors are to be used. It would seem to be best if the constant prior was used in all cases.

Consider again the Poisson example with the noninformative prior $\pi(\theta) = 1$. The estimate is $\hat{\theta} = \overline{x} + 1/t = .5$. The SBC for this model is

$$\begin{aligned} &ln\Pi f(x \mid \theta = .5) + ln(1/\theta) - 1ln(10/2\pi) \\ &\quad = \Sigma ln(e^{-.5}.5^{x_i}/x_i!) - .46471 \\ &\quad = 7(-.5) + 2(-.5 + ln(.5)) + (-.5 + ln(.25/2)) + ln(2.5) - .46471 \\ &\quad = -8.46574 - .46471 = -8.93045. \end{aligned}$$

Now consider the negative binomial model

$$f(x \mid k,p) = \frac{\Gamma(k+x)}{\Gamma(k)\Gamma(x+1)}p^k(1-p)^x, \quad x = 0,\ 1,\ \ldots \tag{2.23}$$

and the prior distribution $\pi(k,p) = 1$. The posterior distribution from a sample of size n is proportional to

$$\frac{\Pi\Gamma(k+x_i)}{\Gamma(k)^n} p^{nk}(1-p)^{\Sigma x_i}. \tag{2.24}$$

For the example this is

$$\frac{\Gamma(k)^7\Gamma(k+1)^2\Gamma(k+2)}{\Gamma(k)^{10}} p^{10k}(1-p)^4 = k^3(k+1)p^{10k}(1-p)^4. \tag{2.25}$$

In this case the noninformative prior will not work; (2.25) cannot be integrated and so is not proportional to a density. While the posterior mean is not available, the posterior mode is. It is the parameter values that maximize (2.25). The maximum occurs at $k = 2.68203$, $p = .87202$. Also note that the joint density required for the SBC is just (2.25) divided by $\Pi\Gamma(x_i + 1) = \Pi x_i! = 2$. So the SBC value is

$$\begin{aligned} &ln[(2.68203)^3(3.68203)(.87202)^{26.8203}(.12798)^4/2] - 2ln(10/2\pi) \\ &\quad = -8.32634 - .92942 = -9.25576. \end{aligned}$$

Without the SBC correction, the negative binomial yields the larger $(-8.32634 > -8.46574)$ value while after the correction the Poisson model emerges as the winner. The increase due to the extra parameter was not sufficient to overcome the estimation error that is introduced by using the same data to estimate two items instead of one. The Poisson model could have been made to look even better had we used the posterior mode (.4) instead of the posterior mean. In that case the SBC value is $-8.35831 - .46471 = -8.82302$.

3. COMPUTATIONAL ASPECTS OF BAYESIAN ANALYSIS

To complete a Bayesian analysis it is often necessary to perform integrations and/or maximizations with respect to functions of many variables. In this Chapter, five approaches will be presented for solving these problems. They all have advantages and disadvantages. Often, but not always, the ones that take the smallest amount of time will be the least accurate. Programs for implementing these procedures are given in the Appendix.

In all cases the problem is to obtain a ratio of integrals of the form

$$\frac{\int g(\boldsymbol{\theta})\pi^*(\boldsymbol{\theta} \mid \boldsymbol{x})d\boldsymbol{\theta}}{\int \pi^*(\boldsymbol{\theta} \mid \boldsymbol{x})d\boldsymbol{\theta}} \tag{3.1}$$

where $\boldsymbol{\theta}$ is the parameter of interest, $g(\boldsymbol{\theta})$ is a function of $\boldsymbol{\theta}$ whose expectation is desired and $\pi^*(\boldsymbol{\theta} \mid \boldsymbol{x})$ is proportional to the posterior density of $\boldsymbol{\theta}$ given $\boldsymbol{x}$.

Before introducing the various methods an example will be introduced that will be used to illustrate each method.

A. AN EXAMPLE

The following example is hypothetical, but does contain enough complexity to provide a useful illustration of the concepts. Consider the following 25 observations generated from a Weibull distribution with parameters $\alpha = .01$ and $\tau = .75$:

1	2	49	57	91	109	119	119	183	198
261	263	317	401	559	586	598	606	658	753
967	978	1188	1284	2468.					

The Weibull density function is

$$f(x \mid \alpha,\tau) = \alpha\tau x^{\tau-1} exp(-\alpha x^{\tau}),\ x > 0,\ \alpha > 0,\ \tau > 0.$$

With the noninformative prior density $\pi(\alpha,\tau) = 1/(\alpha\tau)$ the posterior density is proportional to

$$\pi^*(\alpha,\tau \mid x_1,\ldots,x_n) \propto \alpha^{n-1}\tau^{n-1}\Pi x_i^{\tau-1} exp(-\alpha\Sigma x_i^{\tau}) = f(\alpha,\tau). \tag{3.2}$$

The posterior means of the parameters are

$$E(\alpha \mid x_1,\ldots,x_n) = \frac{\int\int \alpha f(\alpha,\tau)d\alpha d\tau}{\int\int f(\alpha,\tau)d\alpha d\tau},$$

$$E(\tau \mid x_1,\ldots,x_n) = \frac{\int\int \tau f(\alpha,\tau)d\alpha d\tau}{\int\int f(\alpha,\tau)d\alpha d\tau}. \tag{3.3}$$

All three integrands are gamma functions with respect to α and so the inner integrals can be found analytically. They are

$$E(\alpha \mid x_1,\ldots,x_n) = \frac{\int n!\tau^{n-1}\Pi x_i^{\tau-1}(\Sigma x_i^{\tau})^{-n-1}d\tau}{\int (n-1)!\tau^{n-1}\Pi x_i^{\tau-1}(\Sigma x_i^{\tau})^{-n}d\tau},$$

$$E(\tau \mid x_1,\ldots,x_n) = \frac{\int (n-1)!\tau^{n}\Pi x_i^{\tau-1}(\Sigma x_i^{\tau})^{-n}d\tau}{\int (n-1)!\tau^{n-1}\Pi x_i^{\tau-1}(\Sigma x_i^{\tau})^{-n}d\tau}. \tag{3.4}$$

If an additional observation y is to be obtained, the predictive density will be

$$f^*(y \mid x_1,\ldots,x_n) = \frac{\int n!\tau^{n}\Pi x_i^{\tau-1}y^{\tau-1}(\Sigma x_i^{\tau} + y^{\tau})^{-n-1}d\tau}{\int (n-1)!\tau^{n-1}\Pi x_i^{\tau-1}(\Sigma x_i^{\tau})^{-n}d\tau}. \tag{3.5}$$

In the sections that follow, the general problem will be to obtain an integral of the form

$$\int f(\boldsymbol{x})d\boldsymbol{x} \tag{3.6}$$

where the range of integration is arbitrary unless specifically stated.

B. NUMERICAL INTEGRATION

While all the methods in this Chapter might be called numerical integration methods, in this Section the discussion will be confined to linear approaches. In the most general terms that means the approximation takes the form

$$\int f(\boldsymbol{x})d\boldsymbol{x} \doteq \sum_{i=1}^{n} a_i f(\boldsymbol{x}_i) \tag{3.7}$$

where $\boldsymbol{x}_1,\ldots,\boldsymbol{x}_n$ are specially selected arguments and $a_1,\ldots,a_n$ are specially selected weights. These formulas are derived in most all numerical analysis texts and no special discussion in needed here. The major problem is extending these methods to higher dimensional integrals. Begin with a one-dimensional formula:

$$\int f(x)dx \doteq \sum_{i=1}^{n} a_i f(x_i). \tag{3.8}$$

and extend it to k dimensions as

$$\int\cdots\int f(\boldsymbol{x})dx_1\cdots dx_k \doteq \sum_{i_k=1}^{n}\cdots\sum_{i_1=1}^{n} a_{i_1}\cdots a_{i_k} f(x_{i_1},\ldots,x_{i_k}). \tag{3.9}$$

The problems with this approach are obvious. If there are ten terms in the one-dimensional formula, there will be 10^k terms in the k-dimensional formula, many of which will make a negligible contribution to the sum. To use (3.9) we must either obtain an extremely fast computer or a quadrature formula that uses a very small number of terms. Two quadrature formulas will be recommended. The first, adaptive Gaussian integration, requires a large number of function evaluations and is extremely accurate. The second, Gauss-Hermite integration, uses a small number of function evaluations, but is less accurate.

1. Adaptive Gaussian Integration

Gaussian integration is a standard method for approximating an integral over a bounded interval using a specified number of points. The adaptations are designed to first achieve a pre-specified level of accuracy over a bounded interval and then extend the integration to an unbounded interval. This method is given in most all introductory numerical analysis texts (e.g., Burden and Faires, 1989). It is only practical for one dimensional integrals, but is outstanding for that case. The details of this method are presented in the Appendix.

For the Weibull example the two moments are $E(\alpha \mid x_1,\ldots,x_{25}) = .0086617$ and $E(\tau \mid x_1,\ldots,x_{25}) = .83011$. The integrands of (3.4) were evaluated by computing the natural logarithm, adding 240, and then exponentiating. The addition of 240 was done to ensure that the result would be large enough to avoid underflow problems when exponentiated. This will usually be necessary since the integrands tend to involve very small numbers. When the ratios are taken to produce the answer the $exp(240)$ terms will cancel and so will have no effect on the result.

Figure 3.1
Predictive and Fitted Distributions

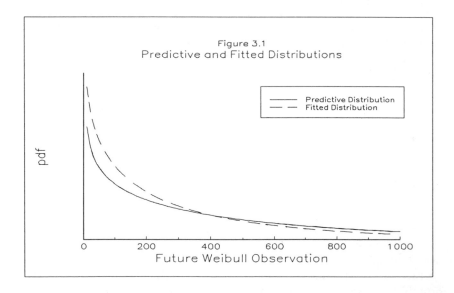

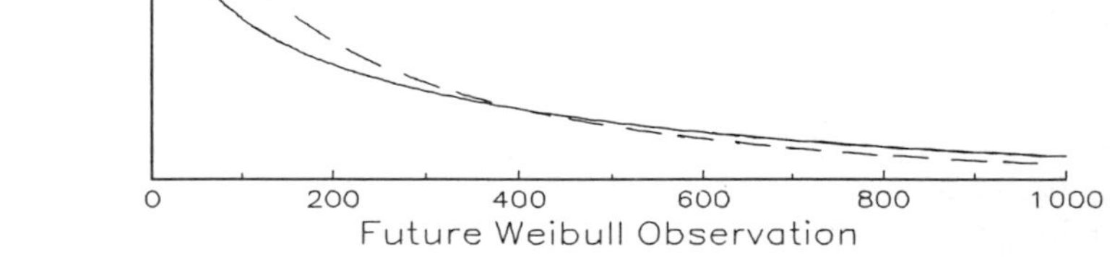

Some selected points from the predictive distribution were also obtained and a graph of this distribution based on 100 points is given in Figure 3.1. This Figure also shows the density function for a Weibull distribution with parameters $\alpha = .0086617$ and $\tau = .83011$. This is the predictive distribution based on the parameter estimates and does not reflect the variability in the prediction due to the estimation of these parameters. As expected, this density does not display as much variation.

It should be noted that the expected value of this predictive distribution does not exist and so the normal approximation is not available. If a prediction interval is desired it will have to be constructed by double integration over τ and y.

2. Gauss-Hermite Integration

This method is specifically designed for integrating over all real numbers. If the range of integration is not the entire real line (as in the case of scale parameters) a transformation, usually the natural logarithm, will be necessary. This is illustrated in the example at the end of this section. An added benefit is that the function is more symmetric, which improves the accuracy of this integration formula. Like most quadrature formulas this one is designed to be exact when the function has a particular form and to be close when the function is close to having this form. For Gauss-Hermite integration the form is a polynomial times a standard normal probability density. The general formula for a one-dimensional integral (as given in Abramowitz and Stegun, 1964, formula 25.4.46) is

$$\int_{-\infty}^{+\infty} g(x)e^{-x^2}dx \doteq \sum_{i=1}^{n} h_i g(x_i). \tag{3.10}$$

where the constants, h_i, and the evaluation points, x_i, are determined so that the formula will be exact when $g(x)$ is a polynomial of degree $2n-1$. Tables of the appropriate constants are given in the Appendix. In this Section all integrals will be assumed to be from $-\infty$ to $+\infty$.

Naylor and Smith (1982) provide the necessary details for implementing (3.10) in the Bayesian framework. The basic problem is that in the multidimensional setting the function multiplying $g(\boldsymbol{x})$ should be approximately multivariate normal with mean zero and identity covariance matrix. This is unlikely to be true, so a transformation is required to achieve this goal. The first step is to generalize the one-dimensional (3.10) to the situation where the mean is μ and the standard deviation is σ. Rewrite the integral as

$$\int f(x)dx = \int f(x)(2\pi\sigma^2)^{1/2}e^{\frac{1}{2}\left(\frac{x-\mu}{\sigma}\right)^2}(2\pi\sigma^2)^{-1/2}e^{-\frac{1}{2}\left(\frac{x-\mu}{\sigma}\right)^2}dx$$

$$= \int g(x)(2\pi\sigma^2)^{-1/2}e^{-\frac{1}{2}\left(\frac{x-\mu}{\sigma}\right)^2}dx. \tag{3.11}$$

Now make the transformation $z = (x-\mu)/\sqrt{2}\sigma$ to obtain

$$\int f(x)dx = \int g(\mu + \sqrt{2}\sigma z)\pi^{-1/2}e^{-z^2}dx. \tag{3.12}$$

The Gauss-Hermite approximation is

$$\int f(x)dx \doteq \sum_{i=1}^{n}\pi^{-1/2}h_i g(\mu + \sqrt{2}\sigma x_i). \tag{3.13}$$

Using the relationship between $g(x)$ and $f(x)$ as given in (3.11) yields

$$\int f(x)dx \doteq \sum_{i=1}^{n}\sqrt{2}\sigma e^{x_i^2}h_i f(\mu + \sqrt{2}\sigma x_i) = \sum_{i=1}^{n} m_i f(z_i)$$

where

$$m_i = \sqrt{2}\sigma e^{x_i^2}h_i, \qquad z_i = \mu + \sqrt{2}\sigma x_i. \tag{3.14}$$

Let us now generalize (3.14) to the multidimensional case. The problem now is to integrate $\int f(\boldsymbol{x})d\boldsymbol{x}$ where all the integrals are from $-\infty$ to $+\infty$. Rewrite (3.11) using a multivariate normal density with mean μ and covariance matrix Σ. We have (with k being the dimension of $\boldsymbol{x}$)

$$\int f(\boldsymbol{x})\mathrm{d}\boldsymbol{x} = \int f(\boldsymbol{x})(2\pi)^{k/2}|\Sigma|^{1/2}exp[(\boldsymbol{x}-\boldsymbol{\mu})'\Sigma^{-1}(\boldsymbol{x}-\boldsymbol{\mu})/2](2\pi)^{-k/2}$$

$$\times|\Sigma|^{-1/2}exp[-(\boldsymbol{x}-\boldsymbol{\mu})'\Sigma^{-1}(\boldsymbol{x}-\boldsymbol{\mu})/2]\mathrm{d}\boldsymbol{x}$$

$$= \int g(\boldsymbol{x})(2\pi)^{-k/2}|\Sigma|^{-1/2}exp[-(\boldsymbol{x}-\boldsymbol{\mu})'\Sigma^{-1}(\boldsymbol{x}-\boldsymbol{\mu})/2]\mathrm{d}\boldsymbol{x}. \tag{3.15}$$

Now let $\Sigma = HDH'$ where D is diagonal and H is lower triangular with ones on the diagonal. If Σ is positive definite the factorization can be done by using the Choleski method. Next, make the transformation

$$\boldsymbol{\eta} = D^{-1/2}H^{-1}(\boldsymbol{x}-\boldsymbol{\mu})/\sqrt{2}.$$

The Jacobian is $2^{k/2}|\Sigma|^{1/2} = 2^{k/2}(d_1\cdots d_k)^{1/2}$ where d_i is the ith diagonal element of D. The integral becomes

$$\int f(\boldsymbol{x})d\boldsymbol{x} = \int g(\boldsymbol{\mu}+\sqrt{2}HD^{1/2}\boldsymbol{\eta})\pi^{-k/2}exp(-\boldsymbol{\eta}'\boldsymbol{\eta})d\boldsymbol{\eta}. \tag{3.16}$$

The multidimensional analog of (3.13) is

$$\int f(\boldsymbol{x})d\boldsymbol{x} \doteq \sum_{i_k=1}^{n}\cdots\sum_{i_1=1}^{n} \pi^{-k/2}h_{i_1}\cdots h_{i_k}g(\boldsymbol{\mu}+\sqrt{2}HD^{1/2}\boldsymbol{x}) \tag{3.17}$$

where $\boldsymbol{x}' = (x_{i_1},\ldots,x_{i_k})$. It is assumed that the number of points, n, used for the approximation is the same for each dimension. The same set of h's and x's are used in each dimension.

Once again, use the definition of $g(\boldsymbol{x})$, this time as in (3.15) to rewrite (3.17) as

$$\int f(\boldsymbol{x})d\boldsymbol{x} \doteq 2^{k/2}\left(\prod_{j=1}^{k} d_j^{1/2}\right)\sum_{i_k=1}^{n}\cdots\sum_{i_1=1}^{n} m_{i_1}\cdots m_{i_k}f(\boldsymbol{z})$$

where

$$m_i = h_i exp(x_i^2) \quad \text{and} \quad \boldsymbol{z} = (\boldsymbol{\mu}+\sqrt{2}HD^{1/2}\boldsymbol{x}). \tag{3.18}$$

Recall from (3.1) that our goal is to integrate functions of the form $g(\boldsymbol{\theta})\pi^*(\boldsymbol{\theta})$. If $g(\boldsymbol{\theta})$ is a polynomial (as would be the case if we are evaluating moments) and $\pi^*(\boldsymbol{\theta})$ is a multivariate normal density (as the Central Limit Theorem says it almost is) then (3.15) indicates that the integration will be exact. This suggests that $\boldsymbol{\mu}$ and Σ be taken as the mean vector and covariance matrix of the posterior distribution. Finding them is the subject of the next subsection.

3. Estimating the Mean and Covariance[5]

Naylor and Smith (1982) suggest an iterative approach to finding $\boldsymbol{\mu}$ and Σ. The idea is to postulate $\boldsymbol{\mu}_0$ and Σ_0 and then use (3.18) to find new approximations, say $\boldsymbol{\mu}_1$ and Σ_1. Continue this process until convergence is obtained. The steps are as follows:

1. Let $\boldsymbol{\mu}_0$ and Σ_0 be the current estimates of the mean and covariance of$\pi^*(\boldsymbol{\theta})$.

2. Write $\Sigma_0 = H_0 D_0 H'_0$ where D_0 is diagonal and H_0 is lower triangular with ones on the diagonal.

3. Let d_j be the jth diagonal element of D_0 and obtain $m_j = h_j exp(x_j^2)$.

4. Use (3.18) to evaluate the following integrals.

   Let M be the result when $f(\boldsymbol{\theta}) = \pi^*(\boldsymbol{\theta})$.
   Let M_i be the result when $f(\boldsymbol{\theta}) = \theta_i \pi^*(\boldsymbol{\theta})$.
   Let M_{ij} be the result when $f(\boldsymbol{\theta}) = \theta_i \theta_j \pi^*(\boldsymbol{\theta})$.

5. Then let $(\boldsymbol{\mu}_1)_i = M_i/M$ and $(\Sigma_1)_{ij} = M_{ij}/M - M_i M_j/M^2$.

6. Let $\boldsymbol{\mu}_0 = \boldsymbol{\mu}_1$ and $\Sigma_0 = \Sigma_1$. Return to step 2 and continue until the values do not change.

Once this is finished, integrals for arbitrary functions $f(\boldsymbol{\theta}) = g(\boldsymbol{\theta})\pi^*(\boldsymbol{\theta})$ can be evaluated using (3.18). The posterior mean and covariance matrix of $\boldsymbol{\theta}$ are already available as $\boldsymbol{\mu}_0$ and Σ_0.

The one problem that remains is the preliminary estimation of the mean and covariance. The easiest choice for the mean is the mode of the posterior density. The mode can be found by applying the simplex method to the logarithm of the posterior density (see the Appendix for a description of the simplex algorithm). The covariance can be approximated by the negative inverse Hessian of the logarithm of the density evaluated at the posterior mode. If $\pi^*(\boldsymbol{\theta})$ is the posterior density for a k-dimensional variable $\boldsymbol{\theta}$, then the ijth element of the Hessian is given by

[5] From here on the word "mean" will always refer to the vector of means and the word "covariance" to the covariance matrix.

$$\left.\frac{\partial^2 ln\pi^*(\boldsymbol{\theta})}{\partial\theta_i\partial\theta_j}\right|_{\boldsymbol{\theta}=\hat{\boldsymbol{\theta}}} \tag{3.19}$$

where $\hat{\boldsymbol{\theta}}$ is the posterior mode. In most cases it will not be possible to analytically obtain the second partial derivatives and so a numerical approximation will be needed. The one suggested in Dennis and Schnabel (1983) is

$$\frac{\pi^*(\boldsymbol{\theta}+c_i\boldsymbol{e}_i+c_j\boldsymbol{e}_j)-\pi^*(\boldsymbol{\theta}+c_i\boldsymbol{e}_i)-\pi^*(\boldsymbol{\theta}+c_j\boldsymbol{e}_j)+\pi^*(\boldsymbol{\theta})}{c_ic_j} \tag{3.20}$$

where $\boldsymbol{e}_i$ is vector with a one in the ith position and zeros elsewhere and the step size c_i is $\zeta^{1/3}\theta_i$ where ζ is the maximum relative error in obtaining values of the function $\pi^*(\boldsymbol{\theta})$. A reasonable default value for ζ is 10^{-d} where d is the number of significant digits used by the computer when doing calculations. The program GHINT in the Appendix calls various subroutines that perform all the steps of a Gauss-Hermite integration. The only input (other than the function to be integrated) is a starting vector for the search for the posterior mode.

4. An Example

The next task is to see how well the Gauss-Hermite method works on the Weibull example. To give this method a full workout, no attempt will be made to reduce the dimension of the problem. The original posterior density is proportional to

$$\pi^*(\alpha,\tau)=\alpha^{n-1}\tau^{n-1}\Pi x_i^{\tau-1}exp(-\alpha\Sigma x_i^{\tau})d\alpha d\tau. \tag{3.21}$$

With the recommended transformation a $= ln(\alpha)$, $t = ln(\tau)$, we have

$$\pi^*(a,t)=exp(an+tn)\Pi x_i^{e^t-1}exp(-e^a\Sigma x_i^{e^t}). \tag{3.22}$$

To indicate the value of the transformation in (3.22), the contours of the standardized (using the mean vector and covariance matrix computed at the end of this Section) density are plotted in Figure 3.2. While not the perfect concentric circles of the standard bivariate normal distribution, this plot does indicate that Gauss-Hermite integration is likely to produce a satisfactory answer.

Figure 3.2
Transformed Posterior Density of a and t

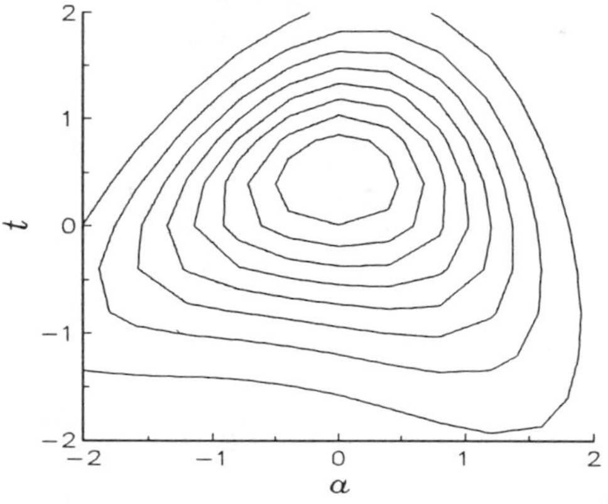

The next step is to get starting values for the mean and covariance of the density in (3.22). The simplex method was used to find the mode at (−5.1,−.18). It was easier to work with the log of (3.22) to avoid underflow problems. It turns out that the value of (3.22) at the maximum is about e^{-180}. To make future evaluations of (3.18) amenable to calculation, it was multiplied by e^{180}. Each time an evaluation was needed the log was obtained, 180 added, and then the result exponentiated. This approach is used in all future evaluations. Because this function is easy to evaluate and thus iterations would be nearly instantaneous, no attempt is made to find a good starting value for the covariance. The following were used:

$$\boldsymbol{\mu}_0 = \begin{bmatrix} -5.1 \\ -.18 \end{bmatrix} \qquad \Sigma_0 = \begin{bmatrix} .85 & -.15 \\ -.15 & .03 \end{bmatrix}.$$

To illustrate the procedure, one iteration with a two point integration formula will be done in great detail. The two points are

$$x_1 = -\sqrt{.5} = -0.7071 \quad x_2 = 0.7071 \quad h_1 = h_2 = 0.8862.$$

The values of m_i, from (3.18) are

$$m_1 = (.8862)exp(.5) = 1.4611 = m_2.$$

The covariance Σ_0 is factored as

$$H_0 = \begin{bmatrix} 1 & 0 \\ -3/17 & 1 \end{bmatrix} \qquad D_0 = \begin{bmatrix} 17/20 & 0 \\ 0 & 3/850 \end{bmatrix}.$$

The four required values of $z = (\boldsymbol{\mu}_0 + \sqrt{2}HD^{1/2}\boldsymbol{x})$ are

i_1	i_2	x_1	x_2	z_1	z_2
1	1	−.7071	−.7071	−6.02195	−.07671
2	1	.7071	−.7071	−4.17805	−.40211
1	2	−.7071	.7071	−6.02195	.04211
2	2	.7071	.7071	−4.17805	−.28329

The values to add for the six integrals that must be evaluated are (since $m_1 = m_2$ they do not need to appear in the calculations, the four points get equal weights)

i_1	i_2	$\pi^*(z)$	$z_1\pi^*$	$z_2\pi^*$	$z_1^2\pi^*$	$z_1z_2\pi^*$	$z_2^2\pi^*$
1	1	.24582	−1.48032	−.01886	8.91439	.11356	.00145
2	1	.26164	−1.09315	−.10521	4.56721	.43956	.04231
1	2	.00553	−.03330	.00023	.20054	−.00140	.00001
2	2	.04429	−.18505	−.01255	.77313	.05242	.00355
Total		.55728	−2.79182	−.13639	14.45527	.60414	.04732
Total/.55728			−5.00973	−.24474	25.93897	1.08409	.08491

Note that the lead coefficient in (3.18) never needs to be calculated since it will cancel when the ratio is taken. The first two terms are the new values for the mean while the new variance terms are $25.93897 - (-5.00973)^2 = .84158$ and $.08491 - (-.24474)^2 = .02426$. The new covariance term is $1.08409 - (-5.00973)(-.24474) = -.14199$. Summarizing:

$$\boldsymbol{\mu}_1 = \begin{bmatrix} -5.00973 \\ -.24474 \end{bmatrix} \qquad \Sigma_1 = \begin{bmatrix} .84158 & -.14199 \\ -.14199 & .02426 \end{bmatrix}.$$

To complete this example, suppose the iterations had ended at this point and the two point integration formula was to be used to evaluate the expected value of α. In the transformed space this would correspond to using $g(a,t) = e^a$. First, obtain the factorization of Σ_1. It is

$$H_1 = \begin{bmatrix} 1 & 0 \\ -.16872 & 1 \end{bmatrix} \qquad D_1 = \begin{bmatrix} .84158 & 0 \\ 0 & .00030 \end{bmatrix}.$$

Next, find the values of z and the sums for the denominator and for the integrand:

i_1	i_2	z_1	z_2	$\pi^*(z)$	$z_1\pi^*(z)$
1	1	−5.92711	−.10728	.14670	.0003911
2	1	−4.09235	−.41684	.24916	.0041610
1	2	−5.92711	−.07264	.43297	.0011544
2	2	−4.09235	−.38220	.35104	.0058624
Total				1.17987	.0115689
Ratio					.00981

When this process was done for real a three point formula was used for the first set of iterations. Once convergence was obtained the process was repeated with the four, five, and then the six point formulas. The idea was to have few iterations required when the number of function calls per iteration is large. In essentially no time at all the following mean and covariance were obtained:

$$\boldsymbol{\mu} = \begin{bmatrix} -5.13657 \\ -.19916 \end{bmatrix} \qquad \Sigma = \begin{bmatrix} .83381 & -.14498 \\ -.14498 & .02690 \end{bmatrix}.$$

These were then used to evaluate the three integrals needed to obtain the same posterior quantities obtained using adaptive Gaussian integration. They are displayed in Table 3.1. In this example the Gauss-Hermite formula provided acceptable answers.

Table 3.1

Weibull Data — Gauss-Hermite Integrals

	Gauss-Hermite integral	*Adaptive integral*	*relative error*
$E(\alpha)$	.0086399	.0086617	.00252
$E(\tau)$	.83038	.83011	.00033

C. MONTE CARLO INTEGRATION

This method of approximate integration attempts to address some of the problems associated with the quadrature formulas presented in Section B. Its main advantage is that it is not dependent on the dimension of the integrand. The basic idea is quite simple. Suppose we want to find $\int f(\boldsymbol{\theta})d\boldsymbol{\theta}$. Consider an arbitrary, absolutely continuous, *pdf*, $h(\boldsymbol{\theta})$, with support that includes the region of integration. Now let $M(\boldsymbol{\theta}) = f(\boldsymbol{\theta})/h(\boldsymbol{\theta})$. Then,

$$E_{h(\boldsymbol{\theta})}[M(\boldsymbol{\theta})] = \int M(\boldsymbol{\theta})h(\boldsymbol{\theta})d\boldsymbol{\theta} = \int f(\boldsymbol{\theta})d\boldsymbol{\theta}. \quad (3.23)$$

Now suppose we are able to obtain a random sample, $\boldsymbol{\theta}_1,\ldots,\boldsymbol{\theta}_n$, from a random variable with density $h(\boldsymbol{\theta})$. Then the strong law of large numbers assures us with probability one that

$$\lim_{n\to\infty} \frac{1}{n}\sum_{i=1}^{n} M(\boldsymbol{\theta}_i) = E_{h(\boldsymbol{\xi})}[M(\boldsymbol{\theta})] = \int f(\boldsymbol{\theta})d\boldsymbol{\theta}. \quad (3.24)$$

As with Gaussian quadrature, any degree of accuracy can be obtained, provided enough points are used. Unlike Gaussian quadrature, it is not so easy to evaluate the error.

There are several issues that must be addressed before this approach can be implemented. The first is the selection of n. Kloek and van Dijk (1978) indicate that in order to be $100(1-p)\%$ confident that the relative error will be less than $100k\%$ the number of observations should be

$$(z_{1-p/2}/k)^2(\sigma^2/\mu^2) \quad (3.25)$$

where $\mu = E[M(\boldsymbol{\theta})]$, $\sigma^2 = Var[M(\boldsymbol{\theta})]$, and z_p is the pth percentile of the standard normal distribution. This formula is identical to those developed in limited fluctuation credibility. Of course μ and σ^2 cannot be known in advance since they are obtained by evaluating the same integrals we are interested in. The way to use (3.25) is to employ a relatively small sample to obtain preliminary estimates and then use these estimates to determine how large the ultimate sample should be. The formula is

$$\frac{\sigma^2}{\mu^2} \doteq \frac{n\Sigma[f(\theta_i)/h(\theta_i)]^2}{[\Sigma f(\theta_i)/h(\theta_i)]^2} - 1. \tag{3.26}$$

A second issue is the selection of the *pdf* $h(\boldsymbol{\theta})$. It should have two properties. First, generation of simulated observations should be easy. Second, it should have a shape that is similar to that of $f(\boldsymbol{\theta})$. The reason for this latter requirement is that such similarity will produce an $M(\boldsymbol{\theta})$ that is nearly constant and will therefore have a small value of σ^2. From (3.25) we see this reduces the number of simulated values needed to achieve a given level of accuracy. For unimodal, somewhat symmetric situations, van Dijk and Kloek (1980) recommend use of the multivariate t distribution. The *pdf* is

$$f(\boldsymbol{\theta} \mid d,\boldsymbol{\mu},A) = \frac{\Gamma((d+k)/2)}{|A|^{1/2}(d\pi)^{k/2}\Gamma(d/2)}\left[1 + \frac{1}{d}(\boldsymbol{\theta} - \boldsymbol{\mu})'A^{-1}(\boldsymbol{\theta} - \boldsymbol{\mu})\right]^{-(d+k)/2} \tag{3.27}$$

where k is the dimension of $\boldsymbol{\theta}$, d is the degrees of freedom, $\boldsymbol{\mu}$ is the mean (if $d > 1$, otherwise it does not exist), and $dA/(d-2)$ is the covariance matrix (if $d > 2$). Since it has larger tails than the normal distribution it is likely to produce more reasonable values of $M(\boldsymbol{\theta})$ for extreme observations. It may be necessary, especially when the components of $\boldsymbol{\theta}$ are variances, to do a log transformation before performing the Monte Carlo integration. This will be illustrated in the examples that appear later in this Section. Generating observations from this distribution is particularly easy. For a $t_k(d,\boldsymbol{\mu},\Sigma)$ density begin by obtaining two vectors, $\boldsymbol{u}' = (u_1,\ldots,u_k)$ and $\boldsymbol{v}' = (v_1,\ldots,v_d)$ where each of the elements is a random observation from a standard normal distribution[6] and the $k+d$ elements are independent. Then

$$\boldsymbol{\theta} = \boldsymbol{\mu} + H\boldsymbol{u}\left(\frac{d}{\boldsymbol{w}'\boldsymbol{w}}\right)^{1/2} \tag{3.28}$$

[6] An algorithm for obtaining standard normal variables from uniform(0,1) random numbers is given in the Appendix. It is assumed that the reader has access to a high quality uniform(0,1) random number generator.

will have the desired distribution where H is any matrix[7] such that $HH' = \Sigma$.

The next task is to select the parameters of the t distribution. As mentioned above the *pdf* should have a shape that is similar to that of $f(\boldsymbol{\theta})$. Reasonable approximations for the mean and covariance matrix of the posterior distribution were given in Section B and can also be used here. If the posterior mode is on a boundary some arbitrary value will need to be selected. Then a small sample can be used to obtain an improved estimate. An appropriate number of degrees of freedom is the number of degrees of freedom that would be associated with a frequentist estimate of the elements of $\boldsymbol{\theta}$. If $\boldsymbol{\theta}$ is multivariate use the smallest degrees of freedom associated with its elements.

The final point is that ordinarily several integrals of the form (3.1) will be desired. They will differ only in the choice of $g(\boldsymbol{\theta})$. It is not practical to generate a new sequence of random numbers for each one and so the following scheme will be the most efficient. Choose the *pdf* $h(\boldsymbol{\theta})$ to closely match the posterior density $\pi^*(\boldsymbol{\theta} \mid \boldsymbol{x})$, generate random observations $\boldsymbol{\theta}_1,\ldots,\boldsymbol{\theta}_n$ from this *pdf*, and then obtain

$$h_i = \pi^*(\boldsymbol{\theta}_i \mid \boldsymbol{x})/h(\boldsymbol{\theta}_i), \qquad i = 1,\ldots,n. \tag{3.29}$$

The estimate of any integral becomes

$$\frac{\int g(\boldsymbol{\theta})\pi^*(\boldsymbol{\theta} \mid \boldsymbol{x})d\boldsymbol{\theta}}{\int \pi^*(\boldsymbol{\theta} \mid \boldsymbol{x})d\boldsymbol{\theta}} \doteq \frac{\sum_{i=1}^{n} h_i g(\boldsymbol{\theta}_i)}{\sum_{i=1}^{n} h_i}. \tag{3.30}$$

Table 3.2

Weibull Data — Monte Carlo Integrals

	Monte Carlo integral	*Adaptive integral*	*relative error*
$E(\alpha)$	.0087355	.0086617	.00852
$E(\tau)$	.83027	.83011	.00019

This method will now be demonstrated on the Weibull example. The log transformation will again be employed so π^* will be as given in

[7]Since Σ must be positive definite, the Choleski factorization is a good choice.

(3.22). A preliminary run with 1000 iterations took about two minutes and revealed that 656,000 iterations would be needed to provide three digit accuracy with fifty percent probability. Instead, only 50,000 iterations were used, taking about one and one-half hours. The results are given in Table 3.2. Almost two digits of accuracy were achieved in the estimate of $E(\alpha)$while three digits were obtained in the estimate of $E(\tau)$ with many fewer iterations than estimated[8].

D. ADJUSTMENTS TO THE POSTERIOR MODE

The final computational method is by far the fastest, at least when only one integral is required. It was developed by Tierney and Kadane (1986) and is based on a Taylor series expansion using the information matrix. The formula is simple:

$$\frac{\int g(\boldsymbol{\theta})\pi^*(\boldsymbol{\theta} \mid \boldsymbol{x})d\boldsymbol{\theta}}{\int \pi^*(\boldsymbol{\theta} \mid \boldsymbol{x})d\boldsymbol{\theta}} \doteq \frac{|\Sigma^*|^{1/2} g(\hat{\boldsymbol{\theta}}^*)\pi^*(\hat{\boldsymbol{\theta}}^* \mid \boldsymbol{x})}{|\Sigma|^{1/2}\pi^*(\hat{\boldsymbol{\theta}} \mid \boldsymbol{x})}. \tag{3.31}$$

Here Σ is the negative inverse Hessian of $ln\pi^*(\boldsymbol{\theta} \mid \boldsymbol{x})$ evaluated at $\hat{\boldsymbol{\theta}}$, the value of $\boldsymbol{\theta}$ that maximizes $ln\pi^*(\boldsymbol{\theta} \mid \boldsymbol{x})$. Σ^* and $\hat{\boldsymbol{\theta}}^*$ are the corresponding values for the function $lng(\boldsymbol{\theta}) + ln\pi^*(\boldsymbol{\theta} \mid \boldsymbol{x})$. The maximizing values can be found by the simplex method while the Hessians can be approximated using (3.20). This process is easy to program and is relatively fast for finding one integral. On the other hand, the values in the numerator of (3.31) must be found from scratch for each integral. This is in contrast to the methods in Sections B and C in which many integrals can be done simultaneously. As a general integration method it should be noted that it can only be applied to the ratio of integrals. The numerator and denominator of (3.31) do not have any meaning by themselves.

Table 3.3

Weibull Data — Tierney-Kadane Integrals

	Tierney-Kadane integral	*Adaptive integral*	*relative error*
$E(\alpha)$	.0086659	.0086617	.00048
$E(\tau)$	.83022	.83011	.00013

[8]It should be noted that the ratio used in (3.26) is for the denominator of (3.30) and so the relative error is for that quantity and not for the ratio of integrals.

This approach was tried on the Weibull data set. The results are given in Table 3.3. The answers were remarkably good and took less than a minute to compute.

E. EMPIRICAL BAYES STYLE APPROXIMATIONS

The term Empirical Bayes has come to be used in a variety of settings with a variety of meanings. For our purposes, it can refer to any method that attempts to shortcut one of the usual Bayesian steps. The two steps that analysts may prefer to eliminate are the choice of parameters for an informative prior distribution and the integrations. The former shortcut will be addressed in the chapters on credibility, while the latter will be discussed here.

Recall that the general integration problem has to do with finding an integral of the form $\int g(\boldsymbol{\theta})\pi^*(\boldsymbol{\theta})d\boldsymbol{\theta}$. An approximation can be obtained by using the first order Taylor series expansion of $g(\boldsymbol{\theta})$ about some point $\hat{\boldsymbol{\theta}}$. It is

$$g(\boldsymbol{\theta}) \doteq g(\hat{\boldsymbol{\theta}}) + (\boldsymbol{\theta} - \hat{\boldsymbol{\theta}})'\nabla g(\hat{\boldsymbol{\theta}}) \tag{3.32}$$

where $\nabla g(\hat{\boldsymbol{\theta}})$ is the vector of partial derivatives of $g(\boldsymbol{\theta})$ taken with respect to each of the elements of $\boldsymbol{\theta}$ and then evaluated at $\hat{\boldsymbol{\theta}}$. Making the substitution in the integral gives

$$\int g(\boldsymbol{\theta})\pi^*(\boldsymbol{\theta})d\boldsymbol{\theta} \doteq g(\hat{\boldsymbol{\theta}}) + (\mu_{\boldsymbol{\theta}} - \hat{\boldsymbol{\theta}})'\nabla g(\hat{\boldsymbol{\theta}}) = g(\mu_{\boldsymbol{\theta}}) \tag{3.33}$$

if $\hat{\boldsymbol{\theta}}$ is taken as $\mu_{\boldsymbol{\theta}}$ where $\mu_{\boldsymbol{\theta}}$ is the posterior mean of $\boldsymbol{\theta}$. This says that, at least approximately, the posterior mean of a function of the parameters can be approximated by evaluating the function at the parameter estimates. One use of this result would be that the predictive distribution can be approximated by inserting the parameter estimates in the density function. It was seen in Figure 3.1 that this may not be a particularly accurate procedure.

The variance of a function of the parameters can also be approximated using this procedure. We have

$$Var[g(\boldsymbol{\theta})] \doteq Var[g(\mu_{\boldsymbol{\theta}}) + (\boldsymbol{\theta} - \mu_{\boldsymbol{\theta}})'\nabla g(\mu_{\boldsymbol{\theta}})] = [\nabla g(\mu_{\boldsymbol{\theta}})]'\Sigma[\nabla g(\mu_{\boldsymbol{\theta}})] \tag{3.34}$$

where Σ is the covariance of the posterior distribution.

These approximations are particularly convenient when the objective is to find the mean and variance of the predictive distribution. For the mean we have

$$E(Y \mid \boldsymbol{x}) = E[E(Y \mid \boldsymbol{\theta},\boldsymbol{x})] = E[m(\boldsymbol{\theta}) \mid \boldsymbol{x}] \doteq m(\mu_{\boldsymbol{\theta}}) \tag{3.35}$$

where $m(\boldsymbol{\theta})$ is the expected value of Y given $\boldsymbol{\theta}$. For the variance we have

$$\begin{aligned} Var(Y \mid \boldsymbol{x}) &= E[\mathrm{Var}(Y \mid \boldsymbol{\theta},\boldsymbol{x})] + Var[E(Y \mid \boldsymbol{\theta},\boldsymbol{x})] \\ &= E[v(\boldsymbol{\theta}) \mid \boldsymbol{x}] + Var[m(\boldsymbol{\theta})] \doteq v(\mu_{\boldsymbol{\theta}}) + [\nabla m(\mu_{\boldsymbol{\theta}})]'\Sigma[\nabla m(\mu_{\boldsymbol{\theta}})] \end{aligned} \tag{3.36}$$

where $v(\boldsymbol{\theta})$ is the variance of Y given $\boldsymbol{\theta}$. The second term reflects the additional variability due to the estimation of $\boldsymbol{\theta}$ while the first term is just the variability in Y itself.

Let us apply these formulas to the Weibull example. Recall that the posterior mean and covariance of the transformed variables a and t were

$$\boldsymbol{\mu} = \begin{bmatrix} -5.13657 \\ -.19916 \end{bmatrix} \qquad \Sigma = \begin{bmatrix} .83381 & -.14498 \\ -.14498 & .02690 \end{bmatrix}.$$

The empirical Bayes style approximation for the mean and variance of $\alpha = exp(a)$ are $E(\alpha) \doteq exp(-5.13657) = .0058778$ and $Var(\alpha) \doteq exp(-10.27314)(.83381) = .000028807$. The values from Gauss-Hermite integration are .0086399 and .000067352 respectively. So here the approximation did not perform so well. The results are bit better for τ, the Gauss-Hermite values are .83038 and .018100 while the empirical Bayes style approximations are .81942 and .018062.

The mean and variance of the predictive distribution were also obtained by both methods. For a Weibull distribution the mean is $\Gamma(1+1/\tau)\alpha^{-1/\tau}$ and the second moment is $\Gamma(1+2/\tau)\alpha^{-2/\tau}$. The Gauss-Hermite integrals with respect to the posterior distribution (using the transformed variables a and t) produced moments of 566.02 and 952,552 and so the variance is $952{,}552 - 566.02^2 = 632{,}173$. The first term of (3.36) is found by inserting the point estimates into the moment formulas to obtain $867{,}438 - 587.99^2 = 521{,}705$. We also see that the estimate of the mean from (3.35) is 587.99. To evaluate the second term of (3.36) we need to evaluate the partial derivatives of $\Gamma(1+e^{-t})exp(-ae^{-t})$ with respect to a and t. They turn out to be −717.57 and −4,012.7. Then the second term of (3.36) is 44,839 for an estimate of the predictive variance of 566,544.

It should be noted that in this example it is unlikely that the empirical Bayes approximation would be used. It requires the posterior mean and variance and these usually have to be obtained by the same approximate integration methods that we are now trying to avoid. Examples will be given in later chapters in which the empirical Bayes approach does indeed obviate the need for approximate integration.

F. SUMMARY

Of the methods presented so far, adaptive integration is clearly the best for one dimension while the Gauss-Hermite approach is best for two to five or six dimensions. For higher dimensions the method of choice is Monte-Carlo integration. The method presented here is capable of significant refinement, but does seem to be a good all-purpose choice. The Tierney-Kadane approach is valuable when just one or two integrals are needed and knowledge of the accuracy of the result is not critical. The empirical Bayes style approximations are available as a last resort. As the power of Bayesian analysis becomes more apparent and its use more widespread the amount of activity in the area of posterior evaluation has been increasing. It is likely that we will soon see even more powerful techniques and commercial computer packages for completing this step.

Shortly after this manuscript was completed a promising new method of high dimensional integration for posterior evaluation was introduced. It is called the Gibbs sampler and a description can be found in Gelfand and Smith (1990). Actuarial examples can be found in Carlin (1991) and Carlin (1992).

4. PREDICTION WITH PARAMETER UNCERTAINTY

A. THE MODEL

The model is the same one that introduced the Bayesian paradigm in Chapter 2. Observations have been obtained from a random variable with known general form, but unknown parameters. Of interest is the value of a future observation whose distribution also depends on these parameters. Of course, this is the traditional actuarial problem. The observations are the benefits paid in the past to policyholders and we desire to predict the payments that will be made in the future.

The first part of this Chapter will deal with a static model in which the universe is unchanging. That is, the process that generates the future losses is the same one that generated the past experience. The only challenge is to identify that process and the only uncertainty is that parameter estimates are used in place of the true values. The latter part of the Chapter will generalize the model to allow the process to be changing over time. Here there is additional uncertainty since the future will not be just like the past. The static model consists of the *pdf* for the observations. In this Chapter we will use $f(\boldsymbol{x} \mid \boldsymbol{\theta})$ to denote this model where $\boldsymbol{x} = (x_1,\ldots,x_n)'$ is the vector of observations and $\boldsymbol{\theta} = (\theta_1,\ldots,\theta_k)'$ is the vector of unknown parameters. Let y be the future observation and let its *pdf* be $g(y \mid \boldsymbol{\theta})$. It is important to note that the density for the future observation need not match that of the past observations. All that is necessary is that it depend on the same parameters.

Once a prior density, $\pi(\boldsymbol{\theta})$, is obtained, it is merely a matter of executing the formulas set out in Chapter 2. Repeating them produces the posterior density

$$\pi^*(\boldsymbol{\theta} \mid \boldsymbol{x}) = \frac{f(\boldsymbol{x} \mid \boldsymbol{\theta})\pi(\boldsymbol{\theta})}{\int f(\boldsymbol{x} \mid \boldsymbol{\theta})\pi(\boldsymbol{\theta})d\boldsymbol{\theta}} \tag{4.1}$$

and the predictive density

$$f^*(y \mid \boldsymbol{x}) = \int g(y \mid \boldsymbol{\theta})\pi^*(\boldsymbol{\theta} \mid \boldsymbol{x})d\boldsymbol{\theta} = \frac{\int g(y \mid \boldsymbol{\theta})f(\boldsymbol{x} \mid \boldsymbol{\theta})\pi(\boldsymbol{\theta})d\boldsymbol{\theta}}{\int f(\boldsymbol{x} \mid \boldsymbol{\theta})\pi(\boldsymbol{\theta})d\boldsymbol{\theta}}. \tag{4.2}$$

The next two sections provide examples using these formulas, one for life insurance and one for casualty insurance. A small example using mortality data for AIDS patients can be found in Klugman (1989). An expanded version of the casualty example is in Klugman (1989).

B. A LIFE INSURANCE EXAMPLE

Historically, life actuarial calculations have proceeded in two steps. The first is to conduct a mortality study of the insured population and from the study estimate the parameters that describe the distribution of the time until death. The second step is to study the future lifetime of a policyholder under the assumption that the parameter estimates are exact. This is the approach taken in the popular actuarial textbooks (e.g., Bowers, et. al. (1986), Parmenter (1988)) and is equivalent to using $g(y \mid \hat{\boldsymbol{\theta}})$ instead of $f^*(y \mid \boldsymbol{x})$. This is precisely the empirical Bayes style approximation from Chapter 3. In this Section an example concerning the evaluation of reserves for a term insurance policy will be discussed.

The policy we will study is a ten year term insurance issued to an individual who is 40 years old. We are particularly interested in the reserve that should be established at age 45, should the policyholder be alive and still insured. To enable us to concentrate on the statistical issues it will be assumed that there are no expenses and that the interest rate is known and fixed at 6%. Then the only uncertain element is when this 45 year old individual will die.

For this example let K be the discrete random variable that measures the complete number of years this individual will live. As this is a term insurance, we can let $K = 5$ if the individual survives to age 50. A mortality investigation has produced the information in Table 4.1 about the distribution of K. Each row of the table represents the result of independent binomial experiments where the sample size is as indicated and the probability of success (death) is θ_i where $\theta_i = Pr($a person age $44 + i$ dies in the next year). These parameters do give the distribution of K as

$$Pr(K=k) = (1-\theta_1)\cdots(1-\theta_k)\theta_{k+1} \text{ for } k = 0,\ldots,4$$

and

$$Pr(K=5) = (1-\theta_1)\cdots(1-\theta_5).$$

The values of $\hat{\theta}_i$ are approximately the ones used to create the 1975-80 Male Ultimate Basic Table (Society of Actuaries, 1985).

Table 4.1

Small Mortality Study

age	*i*	*sample size*	*deaths*	$1000\hat{\theta}_i$
45	1	772	2	2.5907
46	2	1,071	3	2.8011
47	3	1,201	4	3.3306
48	4	1,653	6	3.6298
49	5	1,695	7	4.1298

Finally, assume that calculations are done based on a net annual premium of \$228.83 for a death benefit of \$100,000. The premium is collected at the beginning of the year while the death benefit is paid at the end of the year.

There is one complication that must be discussed first. The usual method for estimating the mortality probabilities is to collect data as in Table 4.1 and find the maximum likelihood (based on the binomial distribution) point estimates. However, it is often the case that the resulting estimates do not follow the smooth, convex, increasing pattern through the middle ages as we know they must. To make the estimates conform to this prior notion they are adjusted by a suitably chosen algorithm. This process is called graduation and an excellent discussion of the various methods can be found in London (1985). One such method is called "the" Bayesian method and does indeed follow the Bayesian paradigm. A second method is called the Whittaker method and London offers a Bayesian rationale for it. As a preliminary step we show how this formula can be derived from the formal Bayesian paradigm and then use it to complete the analysis.

The exact model is that the death counts have a binomial distribution. Approximate this by assuming the mortality estimates $\hat{\theta}_i$ have independent normal distributions with mean θ_i and variance $v_i = \hat{\theta}_i(1-\hat{\theta}_i)/n_i$ where n_i is the sample size. More accuracy could have been achieved by using a variance stabilizing transformation. Doing so would not increase the complexity of the analysis. For the prior distribution assume that the vector $K\boldsymbol{\theta}$ has a multivariate normal distribution with mean $\mathbf{0}$ and covariance $\sigma^2 I$ where K is the matrix that produces the zth differences of a sequence of numbers. For this example we will be using $z = 3$ and so

$$K = \begin{bmatrix} -1 & 3 & -3 & 1 & 0 \\ 0 & -1 & 3 & -3 & 1 \end{bmatrix}. \tag{4.3}$$

This prior is based on an assumption that true values follow a polynomial of degree $z-1$ with some room for variation. To get the prior for $\boldsymbol{\theta}$ assume that K^{-1} exists, in which case $\boldsymbol{\theta}$ has a multivariate normal distribution with mean $\mathbf{0}$ and variance $\sigma^2(K'K)^{-1}$. Of course, this inverse does not exist, which is an indication that this is not a proper prior. One way to eliminate this problem is suggested by Gersch and Kitagawa (1988). Their idea is to fill in additional rows at the top to make K a square matrix. Put small constants in the lower triangle of these extra rows. This creates a proper prior. Letting the constants go to zero creates the answer we are about to obtain.

We now have normal distributions for the model and for the prior. Multiplying them together produces

$$\pi^*(\boldsymbol{\theta} \mid \hat{\boldsymbol{\theta}}) \propto exp[-(\hat{\boldsymbol{\theta}} - \boldsymbol{\theta})'V^{-1}(\hat{\boldsymbol{\theta}} - \boldsymbol{\theta})/2 - \boldsymbol{\theta}'K'K\boldsymbol{\theta}/2] \tag{4.4}$$

where $V = diag(v_1,\ldots,v_5)$. Rearranging the terms, completing the square, and eliminating any terms not involving $\boldsymbol{\theta}$ produces

$$\pi^*(\boldsymbol{\theta} \mid \hat{\boldsymbol{\theta}}) \propto exp[-(\boldsymbol{\theta} - \Sigma V^{-1}\hat{\boldsymbol{\theta}})'\Sigma^{-1}(\boldsymbol{\theta} - \Sigma V^{-1}\hat{\boldsymbol{\theta}})/2],$$

$$\Sigma = (V^{-1} + K'K/\sigma^2)^{-1} \tag{4.5}$$

which is a normal distribution with mean $\tilde{\boldsymbol{\theta}} = \Sigma V^{-1}\hat{\boldsymbol{\theta}}$ and covariance matrix Σ. The posterior mean is the standard Whittaker estimate. The advantage of the Bayesian approach is that we also have the covariance. The usual Whittaker approach is to select a value for σ^2 that produces attractive results. That will be done here, with $\sigma^2 = 10^{-6}$. It should be noted that it is entirely possible to place a prior distribution on σ^2 and then find its Bayes estimate. Such an analysis is best performed using the hierarchical formulas to be introduced in Chapter 7.

For the example we have

$$\tilde{\boldsymbol{\theta}} = (.0025612 \;\; .0028744 \;\; .0032436 \;\; .0036601 \;\; .0041257)'$$

and careful examination will indicate that these values do indeed lie closer to a quadratic function than did the original estimates. The covariance is

$$\Sigma = 10^{-6}\begin{bmatrix} 2.89 & .80 & -.25 & -.40 & .26 \\ .80 & 1.10 & .87 & .35 & -.32 \\ -.25 & .87 & 1.26 & .85 & -.25 \\ -.40 & .35 & .85 & .99 & .54 \\ .26 & -.32 & -.25 & .54 & .22 \end{bmatrix}.$$

The reserve for one such policy is the amount needed to cover the future losses. The random variable is denoted L and is related to K by

$$\begin{aligned} L &= 100{,}000(1.06)^{-K} - 228.83\ddot{a}_{\overline{K+1}|}, \qquad && K = 0,\ 1,\ 2,\ 3, \text{ or } 4 \\ &= -228.83\ddot{a}_{\overline{5}|}, && K = 5. \end{aligned} \tag{4.6}$$

The usual actuarial approach is to obtain the distribution of L using the point estimates from the graduation. It is

k	l	$Pr(L = l)$
0	94,111	.0025612
1	88,555	.0028670
2	83,314	.0032260
3	78,368	.0036285
4	73,704	.0040751
5	– 1,022	.9836522

The net level reserve is the expected value of L, that is, 343.11. Also of interest is the variance of L, 112,890,962.

Now consider the Bayesian prediction of L. From (4.2) we have that the predictive density is

$$Pr(L=l) = \int \prod_{i=1}^{5} (\theta_i)^{r_i}(1-\theta_i)^{s_i}\pi^*(\boldsymbol{\theta} \mid \hat{\boldsymbol{\theta}})d\boldsymbol{\theta}. \tag{4.7}$$

where r_i and s_i are 0 or 1 depending on the value of l and $\pi^*(\boldsymbol{\theta} \mid \hat{\boldsymbol{\theta}})$ is the multivariate normal density with mean $\tilde{\boldsymbol{\theta}}$ and covariance Σ. These integrals could be done analytically, but it is just as easy to use the Gauss-Hermite formula. The predictive distribution is given below. The expected value and variance are 344.63 and 113,017,921. While the change is not much, the variance has increased, as would be expected.

	k	$Pr(L=l)$
0	94,111	.0025641
1	88,555	.0028729
2	83,314	.0032307
3	78,368	.0036317
4	73,704	.0040758
5	− 1,022	.9836248

An interesting aside is the case in which no graduation is done. If the binomial model is used and the prior distribution on a particular probability is $\pi(\theta) \propto \theta^{-1}(1-\theta)^{-1}$, then the predictive distribution and the distribution based the maximum likelihood estimates will be the same. Except for the correlations introduced by the graduation process, the usual actuarial approach is consistent with a Bayesian approach.

C. A CASUALTY INSURANCE EXAMPLE

The problem to be studied here is the classical one of estimating pure premiums. This example concerns the severity component of the pure premium, that is, the amount of a payment given that one has occurred. The typical data set consists of the individual losses that were recorded in each of several prior years. The goal is to find the predictive distribution for the losses in some future year. The particular data used here contained losses on comprehensive general liability policies and was supplied by the Insurance Services Office. The losses were tabulated by accident year (AY) with data from AY 1973 through AY 1986. Only those losses settled at lag 1 (that is, in the same year) are being studied here. To get the pure premium, data from all lags would have to be considered. These issues and more details about the data are available in Klugman (1989). Looking just at lag 1 is sufficient to illustrate a Bayesian analysis. Also, in this Section we assume that the underlying distribution is constant over the various AY's (that is, no inflation, no changes in the tort system or policy forms, etc.). Therefore, the observations can be lumped into one data set. The possibility that the distribution changes over time will be taken up in the next two Sections.

The major difficulty with this data set is that there are 95,231 observations over the 14 AY's. Evaluations of the integrand for any posterior calculations would take much too long. One way to condense the problem is to use the asymptotic properties of maximum likelihood estimators. Suppose the model for a single observation is the density $f(x \mid \boldsymbol{\theta})$. Then the *mle* of $\boldsymbol{\theta}$, $\hat{\boldsymbol{\theta}}$, has an approximate normal distribution with mean $\boldsymbol{\theta}$ and covariance Σ where Σ is the inverse of the information matrix. That is, the ijth element of Σ^{-1} is

$$-nE\left[\frac{\partial^2 ln f(X \mid \boldsymbol{\theta})}{\partial\theta_i \partial\theta_j}\right]. \tag{4.8}$$

This expression will usually involve the unknown parameter $\boldsymbol{\theta}$ and so an approximation will be required. The usual approach is to just insert the *mle*. If the expectation cannot be taken (due to the difficulty of performing the required integral), it can be approximated by

$$-\sum_{i=1}^{n}\frac{\partial^2 ln f(x_i \mid \boldsymbol{\theta})}{\partial\theta_i \partial\theta_j}. \tag{4.9}$$

If the derivative is too difficult to obtain, it can be approximated using (3.20). For the example in this Section, the data were available only in frequency table form so the expectation in (4.8) is available directly from the method of scoring (Hogg and Klugman, 1984). The point of all of this is that the normal density $f(\hat{\boldsymbol{\theta}} \mid \boldsymbol{\theta}, \Sigma)$ can be used as the model in place of the original density.

For the example we use the loglogistic density. This does not give a particularly good fit to the data, but is about as good as can be done with a single density. Much better fits can be obtained with a mixture of densities, but this just increases the number of parameters while adding little to the usefulness of this example. In Klugman (1989) these data are analyzed with a mixture of two Pareto distributions. The loglogistic density and distribution functions are

$$f(x \mid \alpha,\lambda) = \frac{e^{\alpha} e^{\lambda e^{\alpha}} x^{e^{\alpha}-1}}{(e^{\lambda e^{\alpha}} + x^{e^{\alpha}})^2}$$

and

$$F(x \mid \alpha,\lambda) = \frac{x^{e^{\alpha}}}{e^{\lambda e^{\alpha}} + x^{e^{\alpha}}}, \quad -\infty < \alpha,\ \lambda < \infty. \tag{4.10}$$

This parametrization (using the logarithms of the usual parameters) was selected for two reasons. First, it is more likely that these parameters will have a normal distribution and second, with their possible values covering the entire real line, there will be no problems with maximization routines or with Gauss-Hermite integration. The maximum likelihood estimate is $\hat{\boldsymbol{\theta}}' = (\hat{\alpha},\hat{\lambda}) = (.18537,\ 5.5654)$ and the covariance is

$$\Sigma = 10^{-6}\begin{bmatrix} 4.9295 & 2.4878 \\ 2.4878 & 20.2104 \end{bmatrix}.$$

For a prior distribution we take the easy choice, $\pi(\alpha,\lambda) \propto 1$. Then the posterior density is

$$\pi^*(\boldsymbol{\theta} \mid \hat{\boldsymbol{\theta}}) \propto exp[-(\hat{\boldsymbol{\theta}} - \boldsymbol{\theta})'\Sigma^{-1}(\hat{\boldsymbol{\theta}} - \boldsymbol{\theta})/2]. \tag{4.11}$$

This is the same normal distribution, that is, the posterior is normal with mean $\hat{\boldsymbol{\theta}}$ and covariance Σ. The advantage of setting this up as a Bayesian analysis is that we are prepared to find the predictive distribution of the next loss. This is provided from

$$f^*(y \mid \hat{\boldsymbol{\theta}}) = \int\int \frac{e^{\alpha} e^{\lambda e^{\alpha}} x^{e^{\alpha}-1}}{(e^{\lambda e^{\alpha}} + x^{e^{\alpha}})^2}(2\pi)^{-1}|\Sigma|^{-1/2}$$

$$\times\, exp\{-[(\alpha,\lambda) - (\hat{\alpha},\hat{\lambda})]\Sigma^{-1}[(\alpha,\lambda)' - (\hat{\alpha},\hat{\lambda})']/2\}d\alpha d\lambda. \tag{4.12}$$

The predictive density is plotted in Figure 4.1. It turns out that the density obtained by inserting the maximum likelihood estimates in (4.10) is virtually identical. As usual, the real difference is that we now possess information about the variability of the prediction.

There is, however, a small degree of extra variability that is noticeable when we try to obtain the expected value of the next loss. The expected value using (4.10) with the *mles* inserted is 1345.2. The mean of the predictive distribution is found from $\int y f^*(y \mid \hat{\theta}) dy$. Inserting (4.12) and then performing the integral with respect to y first indicates a problem. For $\alpha \leq 0$ the integral does not exist! And since our prior admits the possibility that α has a small value our posterior opinion is that this is still possible and so the predictive mean does not exist. Apparently there is a slightly thicker tail in the predictive density. There are two solutions to this problem. One is that perhaps we believe negative values of α are impossible. This is certainly a legitimate belief if we are absolutely sure the population does have a mean. Then the integral with respect to y can always be performed and the integral with respect to α will run from zero to infinity. A denominator for (4.11) and (4.12) will be needed to recognize that the posterior probability is being concentrated in a smaller area.

Figure 4.1
Predictive Distribution

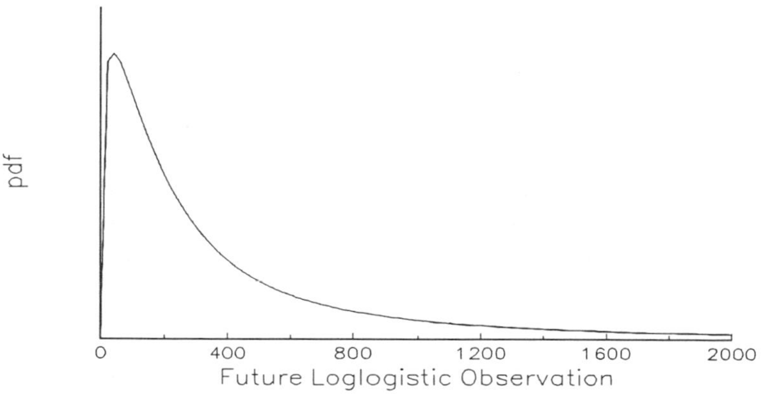

A second remedy is to recognize a potential upper limit for the loss. The integral with respect to y from zero to some finite number will always exist and then the analysis can be completed. This approach is taken here. Suppose there is a policy limit of 100,000. The integral with respect to y can be transformed to be an incomplete beta function and then done first. Gauss-Hermite integration is easy to perform since the posterior distribution is multivariate normal with known mean and variance. The predicted expected value turned out to be 963.19, which is the same value that is obtained when using the maximum likelihood values directly. The variance of the next loss can be found by integrating the square of the loss (another incomplete beta function) and then integrating with respect to the posterior density of the parameters and subtracting the square of the mean. The result is a standard deviation of 4,270.24, which is slightly larger than 4,269.75, reflecting the slight additional variability due to parameter estimation. If a limit of 1,000,000 is used the means are 1,106.17 (Bayesian) and 1,106.15 (maximum likelihood) and the standard deviations are 10,983.91 and 10,981.87 respectively. It is clear that the extra variability in the posterior distribution is in the extreme values.

This example indicates that at times the extra work required for a Bayesian analysis may produce very little reward. In the next section a model that is more realistic will be introduced and it is one that demands a Bayesian solution. In the following section we will return to this particular example and re-analyze it with this new model.

D. THE KALMAN FILTER

Recall that in the example in the previous section the observations were taken over several years and the objective was to predict values that would be observed in future years. It is almost certainly the case that the underlying probability distribution was changing during the observation period and will continue to change in the future. This is another kind of uncertainty and should often be included when constructing models. One extremely useful model is the Kalman filter. Its main features are that it allows for the specific incorporation of time-varying parameters and that no integrations are required to obtain the posterior distribution. Its major drawback is that it accomplishes this by restricting attention to the normal distribution.

The Kalman filter model consists of two parts. The first is the observation equation, which indicates what an observation at time t might be. That is, we have

$$\boldsymbol{X}_t \sim N(A_t\boldsymbol{\theta}_t, F_t). \tag{4.13}$$

The parameter $\boldsymbol{\theta}_t$ is unknown while the covariance matrix F_t may or may not be known. The design matrix A_t must be known. While all three items are subscripted, it is not necessary that they differ from time to time. The usual use for this model is in a linear regression setting where A_t is a design matrix and $\boldsymbol{\theta}_t$ is the regression coefficients. In this Chapter we will use this model as in the previous Section; the data will be reduced to be maximum likelihood parameter estimates. In Chapter 7 another version of this model will be introduced. So rather than a vector of observations at time t, we have the vector of maximum likelihood estimates. Its mean is the true vector of parameters while the variance is the inverse of the information matrix, which will have been estimated from the original data. The observation equation becomes

$$\hat{\boldsymbol{\theta}}_t \sim N(\boldsymbol{\theta}_t, F_t). \tag{4.14}$$

In some of the special cases that follow there will be additional parameters that must be carried from year to year. To accommodate this make a slight change. Introduce the vector

$$\boldsymbol{\mu}_t = \begin{bmatrix} \boldsymbol{\theta}_t \\ \boldsymbol{\beta}_t \end{bmatrix} \tag{4.15}$$

and the matrix

$$A = \begin{bmatrix} I & 0 \end{bmatrix}. \tag{4.16}$$

Then (4.14) becomes

$$\hat{\boldsymbol{\theta}}_t \sim N(A\boldsymbol{\mu}_t, F_t). \tag{4.17}$$

The second part of the model is also expressed as a multivariate normal distribution:

$$\boldsymbol{\mu}_t \sim N(B_t\boldsymbol{\mu}_{t-1}, G_t). \tag{4.18}$$

This formulation indicates how the parameters change from year to year. The design matrix B_t must be known, while the covariance matrix G_t may be unknown.

The standard approach for estimating the parameters is to use the equations associated with the Kalman filter. A good introduction and a derivation of the formulas is found in Meinhold and Singpurwalla (1983) and an actuarial application in deJong and Zehnwirth (1983). Start with initial vector $\tilde{\mu}_0$ and matrix C_0 (these specify a multivariate normal prior distribution) and then update these quantities via

$$C1_t = B_t C_{t-1} B_t' + G_t \tag{4.19}$$

$$V_t = (AC1_t A' + F_t)^{-1} \tag{4.20}$$

$$K_t = C1_t A' V_t \tag{4.21}$$

$$C_t = C1_t - K_t AC1_t \tag{4.22}$$

$$\tilde{\mu}_t = B_t \tilde{\mu}_{t-1} + K_t(\hat{\theta}_t - AB_t \tilde{\mu}_{t-1}). \tag{4.23}$$

These equations can be interpreted as follows: C_{t-1} represents our uncertainty about μ_{t-1} based on all the information we have up to time $t-1$. Using (4.18) to predict μ_t would have us use $B_t\tilde{\mu}_{t-1}$ and $C1_t$ as given by (4.19) expresses the variance of this prediction. We next collect data at time t, which is $\hat{\theta}_t$ with distribution as in (4.17). The unconditional variance of this observation is V_t^{-1}, which consists of two components. The first one in (4.20) is the variance of the changing parameters and the second is the variance of the latest observation. The factor K_t given in (4.21) is a weighting factor used in (4.23) to effect a compromise between the previous estimate and the information from the current observation (as is typical of a Bayes estimate). The variance of the estimate is updated in (4.22), the subtraction indicating that the observation at time t serves to reduce our uncertainty.

At time t the posterior distribution of the parameter μ_t is the normal distribution with mean $\tilde{\mu}_t$ and covariance C_t. Let n be the last year for which data are available and suppose we want to make a prediction about year $n+r$. The predictive distribution for θ_{n+r} is normal with mean and covariance

$$\tilde{\theta}_{n+r} = AB_r^* \tilde{\mu}_n \tag{4.24}$$

$$H_{n+r} = AB_r^* C_n B_r^{*\prime} A' + A(G_{n+r} + B_1^* G_{n+r-1} B_1^{*\prime} + \cdots + B_{r-1}^* G_{n+1} B_{r-1}^{*\prime})A' \tag{4.25}$$

where

$$B_i^* = B_{n+r} B_{n+r-1} \cdots B_{n+r-i+1}. \tag{4.26}$$

If there are any unknown parameters it is customary to estimate them by maximum likelihood[9]. The loglikelihood function is

$$l = \sum_{t=a}^{n} ln|V_t| - \sum_{t=a}^{n} (\hat{\boldsymbol{\theta}}_t - B_t\tilde{\boldsymbol{\mu}}_{t-1})' V_t (\hat{\boldsymbol{\theta}}_t - B_t\tilde{\boldsymbol{\mu}}_{t-1}). \tag{4.27}$$

If a proper prior is being used (C_0 positive definite) the summations can start at $a = 1$. For a noninformative prior use $\tilde{\boldsymbol{\mu}}_0 = \mathbf{0}$ and $C_0 = mI$ where m is a very large number. The value of a must be large enough to allow the process to reach a state where there is enough information to estimate the parameters. In most cases $a = 2$ is appropriate as the model will involve just one lag. That is, two observations will be necessary to get initial estimates.

If no observations are available for one of the time periods, set $V_t = 0$ in (4.20). This will make $K_t = 0$ and so the variance will increase (due to the passage of time) but not be reduced (due to the lack of data). As well, there will be no adjustment in (4.23) reflecting the unavailability of data on which to make one. When evaluating (4.27) the term for that time period should be left out of both sums.

It is likely that the quantity of interest is not the parameters themselves, but some function of them. This can be done as in the last Section, integrating the function of interest against the predictive distribution. Gauss-Hermite integration is likely to do very well as the posterior already has the normal distribution.

A particular predictive case of interest is an observation at time $n + r$. Let the mean of the next observation be $m_1(\boldsymbol{\theta})$, the second moment be $m_2(\boldsymbol{\theta})$, and the variance be $v(\boldsymbol{\theta})$. Then given the data, the posterior values are

$$m_1 = \int m_1(\boldsymbol{\theta})p(\boldsymbol{\theta})d\boldsymbol{\theta}, \quad m_2 = \int m_2(\boldsymbol{\theta})p(\boldsymbol{\theta})d\boldsymbol{\theta}, \quad v = m_2 - m_1^2. \tag{4.28}$$

Note that the unconditional variance cannot be found by integrating $v(\boldsymbol{\theta})$. An empirical Bayes style approximation uses mean $m_1(\tilde{\boldsymbol{\theta}}_{n+r})$ and variance

$$v(\tilde{\boldsymbol{\theta}}_{n+r}) + [\nabla m_1(\tilde{\boldsymbol{\theta}}_{n+r})]' H_{n+r} [\nabla m_1(\tilde{\boldsymbol{\theta}}_{n+r})]. \tag{4.29}$$

---

[9] In this situation the maximum likelihood estimate can be considered as a Bayes estimate based on the posterior mode and a uniform prior. Custom also dictates that these estimates be inserted in (4.19)–(4.23) to produce the posterior distribution of $\boldsymbol{\mu}_n$. This is an empirical Bayes style approximation that eliminates the need for integrating with respect to these "nuisance" parameters.

E. RETURN OF THE CASUALTY INSURANCE EXAMPLE

The example in Section C comprised losses from fourteen accident years. It is reasonable to assume that the parameters of the loglogistic distribution are not constant over that period. At a minimum, inflation would increase the values from year to year. A uniform increase at all loss levels would be reflected in the scale parameter, $exp(\lambda)$, with the shape parameter, α, unchanging. To check this out maximum likelihood estimates were found for each of the fourteen accident years separately along with the estimates of the covariance. The results appear in Table 4.2 and show precisely this pattern. The shape estimates vary about the value 0.21 while the logarithms of the scale parameter are steadily increasing. Exponentiating the differences produces the growth rate. Over the thirteen year period the increase averaged $(6.01930 - 4.80707)/13 = .093248$. Exponentiating gives 1.0977 for an average inflation rate of 9.77%. The decreasing variances over most of the fourteen year period are due to the increasing sample sizes.

Table 4.2

Maximum Likelihood Estimates by Accident Year

AY	$\hat{\alpha}$	$\hat{\lambda}$	$10^5\, Var(\hat{\alpha})$	$10^5\, Cov(\hat{\alpha},\hat{\lambda})$	$10^5\, Var(\hat{\lambda})$
1	0.22875	4.80707	21.75809	6.66079	63.3878
2	0.22594	4.97911	20.18914	6.45544	63.1064
3	0.20461	5.04233	12.92652	4.63235	43.7639
4	0.22545	5.08108	13.42460	4.45479	43.4881
5	0.21740	5.22785	11.96369	4.39330	41.4465
6	0.25411	5.33415	10.40094	3.44992	33.6814
7	0.21872	5.44873	8.56861	3.53802	31.3911
8	0.20267	5.53316	5.35969	2.48265	20.8769
9	0.18805	5.55208	4.78263	2.37081	19.4208
10	0.19164	5.73094	4.43767	2.40954	18.4698
11	0.27965	5.75153	4.37360	1.72644	14.6406
12	0.18762	5.84150	4.61574	2.70918	19.7311
13	0.21088	5.89134	4.88891	2.72800	19.8863
14	0.18568	6.01930	6.27884	4.08937	27.5124

The next step is to devise a Kalman filter model that describes this inflationary process. All of the models have the same observation equation

$$\hat{\boldsymbol{\theta}}_t \sim N(A\boldsymbol{\mu}_t,\, F_t) \tag{4.30}$$

where $\boldsymbol{\mu}_t$ has α_t and λ_t as its first two elements and perhaps some additional elements while $\hat{\boldsymbol{\theta}}_t$ has $\hat{\alpha}_t$ and $\hat{\lambda}_t$ as its only two elements. The matrix A is rectangular with two rows and ones on the main diagonal. The covariance F_t contains the estimates as given in Table 4.2. The second equation describes how the elements of $\boldsymbol{\mu}$ change over time. We begin with the most complex version and then note the various simplifications that can be made. Let (4.18) be written as

$$\boldsymbol{\mu}_t \sim N(B\boldsymbol{\mu}_{t-1}, G). \tag{4.31}$$

Note that the matrices B and G do not change over time. Let $\boldsymbol{\mu}_t$ contain the six elements α_t, λ_t, ρ_t, $\overline{\alpha}_t$, $\overline{\lambda}_t$, and $\overline{\rho}_t$. The parameter ρ_t is the increase in λ_t due to inflation and the three values with the bars on top represent the long-term "averages" of these values. The matrix B has the form given below while the covariance G is zeros except for the first three diagonal elements, which we will label g_1, g_2, and g_3.

$$B = \begin{bmatrix} z_1 & 0 & 0 & 1-z_1 & 0 & 0 \\ 0 & z_2 & 1 & 0 & 1-z_2 & 0 \\ 0 & 0 & z_3 & 0 & 0 & 1-z_3 \\ 0 & 0 & 0 & 1 & 0 & 0 \\ 0 & 0 & 1 & 0 & 1 & 0 \\ 0 & 0 & 0 & 0 & 0 & 1 \end{bmatrix}. \tag{4.32}$$

The first and fourth rows indicate how the shape parameter changes over time. The pattern is

$$E(\alpha_t) = z_1\alpha_{t-1} + (1-z_1)\overline{\alpha}_{t-1} \text{ and } \overline{\alpha}_t = \overline{\alpha}_{t-1}. \tag{4.33}$$

The first equation indicates an autoregressive process of order one. The new value depends partly on the old value and partly on the long-term average plus some noise has governed by the variance g_1. The second equation indicates that the long-term average does not change over time. From the data in Table 4.2 it appears that a simpler model with $z_1 = 0$ may be sufficient. This would indicate that each year's shape parameter is a random deviation from the long-term average but bears no relationship to last year's value. A further simplification is obtained by setting $g_1 = 0$.

This implies that the shape parameter is unchanging over time and the differences in Table 4.2 are due to sampling variation and not any underlying variability in the parameter. Since the covariances in Table 4.2 represent the underlying variability in the data we should be able to test the hypothesis that $g_1 = 0$.

The second and fifth equations are

$$E(\lambda_t) = z_2\lambda_{t-1} + (1 - z_2)\overline{\lambda}_{t-1} + \rho_{t-1} \text{ and } \overline{\lambda}_t = \overline{\lambda}_{t-1} + \rho_{t-1}. \tag{4.34}$$

The second equation is again autoregressive of order one with the addition of the inflation factor ρ_{t-1}. The long-term average in the fifth equation is not really an average, but indicates that the base value is also to be increased by inflation. It is not clear from Table 4.2 if any simplifications are appropriate.

The third and sixth equations indicate how the inflation factor changes over time. They are just like equations one and four with an autoregressive process for the inflation factor and an unchanging long-term average. Again, it is not clear what simplifications, if any, should be used.

There are up to six parameters that must be estimated by maximum likelihood (three z's and three g's) using the likelihood function (4.27) with $a = 3$.[10] The value of the likelihood function at the maximum indicates which model is best. Table 4.3 gives the maximum likelihood estimates, the loglikelihood value, and the Schwartz Bayesian Criterion[11] for a variety of models. In all cases the noninformative prior was created by using an initial covariance of $C_0 = 10^6 I$. Model six was also run using $C_0 = 10^7 I$ and the results were identical to four significant digits.

At model three the autoregressive parameter z_1 is ever so slightly significant. A comparison of models six and eight (both of which are much better than model three) shows that z_1 does not belong in the final model. Of the remaining models it appears that model six is the most reasonable choice. This model indicates that all three parameters are fluctuating

[10] The autoregressive model for the inflation factor requires two observations to provide a point estimate. It requires two observations of the scale parameter to estimate the inflation factor, so it takes three observations to produce an initial estimate from a noninformative prior.

[11] The value of p used here is the number of z's and g's being estimated. The number of observations is 14, the number of maximum likelihood estimates of the loglogistic parameters.

randomly about a long-term value, but there is no dependence on the previous year. For comparison, the analyses that follow will also be done with model one, an inflation model with unchanging parameter values.

Table 4.3

Parameter Estimates for the Kalman Filter Model

model	z_1	z_2	z_3	$10^4 g_1$	$10^4 g_2$	$10^4 g_3$	$ln(l)$	SBC
1	0.0	0.0	0.0	0.0	0.0	0.0	−56.077	−56.077
2	0.0	0.0	0.0	7.8290	0.0	0.0	95.339	94.538
3	−0.3286	0.0	0.0	6.7392	0.0	0.0	96.276	94.674
4	0.0	0.0	0.0	7.4816	0.0	19.4401	127.809	126.207
5	0.0	0.0	−0.0001	7.4845	0.0	17.9491	128.073	125.669
6	0.0	0.0	0.0	7.4110	8.2346	3.0541	130.332	127.928
7	0.0	0.00004	0.0	7.4071	7.8338	2.4879	130.608	127.403
8	−0.2714	0.0	0.0	6.7283	8.0693	3.0489	130.703	127.498

For model one the final parameter estimates are

$$\tilde{\mu}_{14} = (.21253 \quad 6.02885 \quad .08920 \quad .21253 \quad 6.02885 \quad .08920)'$$

$$10^6 C_{14} = \begin{bmatrix} 5.06 & 2.99 & 0.12 & 5.06 & 2.99 & 0.12 \\ 2.99 & 56.96 & 7.43 & 2.99 & 56.96 & 7.43 \\ 0.12 & 7.43 & 1.46 & 0.12 & 7.43 & 1.46 \\ 5.06 & 2.99 & 0.12 & 5.06 & 2.99 & 0.12 \\ 2.99 & 56.96 & 7.43 & 2.99 & 56.96 & 7.43 \\ 0.12 & 7.43 & 1.46 & 0.12 & 7.43 & 1.46 \end{bmatrix}.$$

With no variation in the model, the current values at time fourteen are identical to the estimated long-term averages. For model six the values are

$$\tilde{\mu}_{14} = (.18783 \quad 6.01954 \quad .09000 \quad .21537 \quad 6.01571 \quad .09000)'$$

and

$$10^6 C_{14} = \begin{bmatrix} 57.54 & 32.83 & 1.52 & 4.62 & 17.43 & 1.52 \\ 32.83 & 236.31 & 11.04 & 2.82 & 125.10 & 11.04 \\ 1.52 & 11.04 & 336.39 & -.12 & 44.37 & 30.98 \\ 4.62 & 2.82 & -.12 & 59.51 & 2.19 & -.12 \\ 17.43 & 125.10 & 44.37 & 2.19 & 502.06 & 44.37 \\ 1.52 & 11.04 & 30.98 & -.12 & 44.37 & 30.98 \end{bmatrix}.$$

It is not surprising that the estimates of the long-term average are similar for the two models. The major difference is in the covariances. Model six incorporates more variability; it should show up when we evaluate the predictive distribution for a future period.

Now suppose we are interested in the distribution of losses from accident year 16. Using (4.24) and (4.25) to make the projections, we have for model one

$$\tilde{\boldsymbol{\theta}}_{16} = (.21253 \quad 6.2072)'$$

$$10^6 H_{16} = \begin{bmatrix} 5.0617 & 3.2319 \\ 3.2319 & 92.4955 \end{bmatrix}$$

and for model six

$$\tilde{\boldsymbol{\theta}}_{16} = (.21537 \quad 6.1957)'$$

$$10^6 H_{16} = \begin{bmatrix} 800.617 & 1.9401 \\ 1.9401 & 2238.70 \end{bmatrix}.$$

As expected, the estimates of the two parameters are essentially the same, but model six indicates considerably greater variation. Our discovery that the parameters vary from year to year indicates that predictions may be unreliable.

The last step is to determine the predictive distribution. The two densities are plotted in Figure 4.2. The integrals were evaluated using Gauss-Hermite integration and formula (4.12). The two models appear to produce the same predictive distributions. Although there is greater uncertainty with model six, the variances in H_{16} are still extremely small relative to the parameter values. The statistical significance of model six does not translate into operational significance. In Chapter 7 some more aspects of this problem will be analyzed and perhaps the value of modeling the extra variation will be more apparent. Any moments that are desired must be found as in Section C as the problem of infinite integrals still exists.

Figure 4.2
Predictive Distributions, Models 1 and 6

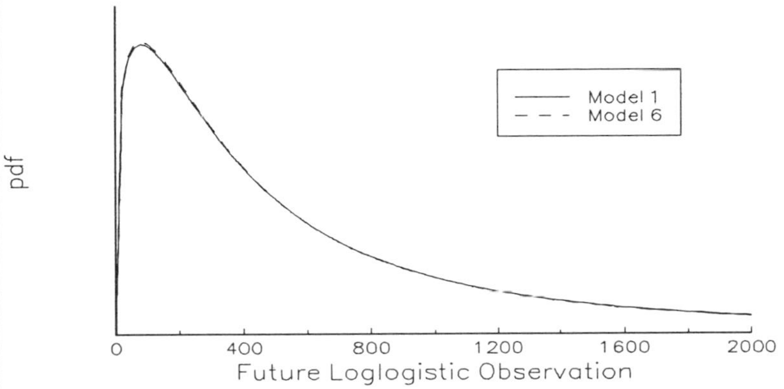

5. THE CREDIBILITY PROBLEM

The problem is estimation of the amount or number of claims to be paid on a particular insurance policy in a future coverage period. This is a random quantity whose ultimate value will be affected by a number of factors: the individual characteristics of the insured, the characteristics of a larger group to which the insured belongs, external factors (mostly economic quantities), and the random nature of the insured event. Recognizing that no amount of information will allow us to exactly predict future claims, we settle for either the probability distribution of this amount or properties of this distribution such as the mean and variance. Of greatest interest is the mean, which (under squared error loss) would be our best guess as to what the future claims might be. For the most part we will ignore the economic variables, or equivalently, assume they are accounted for outside the credibility analysis.

To create the credibility environment we must be concerned with the characteristics that distinguish one insured from another. These are the classification variables that are part of any underwriting process. For this introduction we need not be concerned with what they are, only that the insureds can be partitioned into k classes based on observable characteristics. This broad setting can accommodate both credibility environments — classification rate making and experience rating. In the former problem there are many similar individuals in each group. For example, in automobile insurance a class would be those with certain age–sex–location–marital status values. Each class must be assigned a premium that is consistent with its propensity to produce claims. In the experience rating setting each class is one insured (of course, the one insured could be an employer with many individuals covered under its group policy). Again the objective is to determine the premium for members of the class. It is also possible that both settings exist at once — a driver could be experience rated within the underwriting class to which he has been classified. Over the years the word credibility has come to mean different things to different people, usually based on the restrictions that are to be placed on the solution. What all credibility procedures have in common is the problem to be solved:

> *On the basis of observations from both members of the class under consideration and members of the other classes, estimate the distribution (or its properties) of future claims from members of the class.*

A. A SIMPLE MODEL

To elaborate on the concepts introduced above some notation will be introduced to describe the typical situation. Let there be k classes of policyholders and let there be t observations from each class. Let X_{ij} be the jth observation from class i. Further assume that the probability distribution of X_{ij} depends on a (possibly vector) parameter θ_i. Also assume that given the parameters $\theta_1,\ldots,\theta_k$ the observations are independent. In practice the observations might be loss ratios from each class measured over t years or might be actual losses from t individuals in each class. In either case we want to learn about $X_{i,t+1}$, the next observation in class i.[12] The expected value is

$$E(X_{i,t+1} \mid \theta_i) = m(\theta_i).$$

Our goal is to use the tk observations to obtain an estimate of this quantity.

B. ESTIMATING THE CLASS MEAN

A logical first choice for an estimate of $m(\theta_i)$ is

$$\overline{X}_i = \sum_{j=1}^{t} X_{ij}/t.$$

It is unbiased and enjoys the other standard properties of the sample mean.

[12] For static models, estimating the distribution of the next observation is the same as estimating any future observation, regardless of how far in the future it might be. When we return to the Kalman filter in Chapter 7 the distance into the future becomes relevant.

With no additional knowledge it would be hard to do better. Now suppose we know that the parameters $\theta_1,\ldots,\theta_k$ are independently distributed among the k classes according to some probability distribution that depends on a (possibly vector) parameter μ. There are two ways to think about this second level distribution. One view is that there is a real, but unseen, process that has allocated these parameters to the classes. Learning about this allocation process is just another estimation problem. The second is that this second level distribution represents our knowledge of the classes before collecting the observations. As long as the second level distribution is the same these two approaches will lead to the same result. Those who take the second approach often call their analysis Bayesian, but this is correct only if their knowledge is formulated before collecting the data. More will said about this in the next Chapter. For now the key point is that the classes are linked by the process that generated the class parameters.

If we hope to improve upon the sample mean as a point estimate of the next observation we will need a criterion for evaluating estimates. The measure of choice in the development of credibility theory has been mean squared error. To make this precise, let X without a subscript stand for the entire collection of observations and let $\delta(X)$ be the estimate of $m(\theta_i)$ based on these observations. The mean squared error of the estimate is

$$mse[\delta(X)] = E_\theta\Big[E_{X\mid\theta}\{[\delta(X) - m(\theta_i)]^2\}\Big] \tag{5.1}$$

where the expectation is taken over all possible values of $\theta_1,\ldots,\theta_k$ and of X. For $\delta(X) = \overline{X}_i$ the inner expectation is just the variance of the sample mean which is one over t times the variance of a single observation, $Var(X_{ij}\mid\theta_i) = s(\theta_i)$. The mean squared error is then $E_\theta[s(\theta_i)]/t$, which is a function of the second level parameter μ. Is this the best that can be done? Not necessarily. The above problem is a classical Bayes decision problem as introduced in Chapter 2 and it is not difficult to see (Berger(1985), Chapter 4) that the minimum mean squared error is achieved by taking

$$\delta(X) = E_{\theta\mid X}[m(\theta_i)\mid X], \tag{5.2}$$

the posterior mean of the quantity of interest. It will be a function of the data and of the second level parameter μ. To evaluate this estimate we would not only need to know this parameter but also the form of the distributions of X_{ij} and θ_i.

Now consider the following compromise. Let

$$\delta_C(X) = Z\overline{X}_i + (1 - Z)m \tag{5.3}$$

where $m = E_\theta[m(\theta_i)]$. Note that m is our best guess at the average value from class i prior to having collected any data and is a function of μ. The inner expectation for the mean squared error is

$$\begin{aligned}
&E_{X\mid\theta}\{Z^2\overline{X}_i^2 + 2Z\overline{X}_i[m - Zm - m(\theta_i)] \\
&\quad + (1-Z)^2m^2 + m(\theta_i)^2 - 2(1-Z)mm(\theta_i)\} \\
&\quad = Z^2[s(\theta_i)/t + m(\theta_i)^2] + 2Zm(\theta_i)[m - Zm - m(\theta_i)] \\
&\qquad + (1-Z)^2m^2 + m(\theta_i)^2 - 2(1-Z)mm(\theta_i) \\
&\quad = Z^2s(\theta_i)/t + (1-Z)^2[m(\theta_i) - m]^2.
\end{aligned} \tag{5.4}$$

Now let

$$s = E_\theta[s(\theta_i)] \text{ and } v = Var_\theta[m(\theta_i)] = E\{[m(\theta_i) - m]^2\} = E_\theta[m(\theta_i)^2] - m^2.$$

Then the outer expectation in (5.1) yields

$$mse[\delta_C(X)] = sZ^2/t + (1-Z)^2v. \tag{5.5}$$

Taking the derivative with respect to Z and setting it equal to zero yields

$$Z = \frac{v}{v + s/t}. \tag{5.6}$$

Inserting this back into (5.5) produces

$$mse[\delta_C(X)] = \frac{sv/t}{v + s/t}. \tag{5.7}$$

Since both v and s must be non-negative the *mse* is less than s/t which was the *mse* of the sample mean.

The result presented above had its genesis in the work of Whitney (1918) and the argument given above mirrors Bühlmann (1967). Today, results like this come under the general heading of greatest accuracy or linear Bayes credibility.[13] There are a number of conclusions that can be drawn from this analysis:

1. When class to class relationships are added into the analysis the sample mean is no longer the best estimate.

2. The amount of weight to put on the sample mean is a function of v, s, and t. As t increases so does the weight. This reflects the increased reliability of the sample from the class in question. As s increases the weight decreases. This reflects the increased variability of the observations within a class, indicating the unreliability of the sample mean. As v increases the weight increases. This reflects the increased variability from one group mean to the next. As the other groups become more unlike the group in question their contribution should be reduced.

3. The credibility solution is the best linear approximation to the Bayes solution of using the posterior mean. The advantage of the linear approximation is that only the first two moments of the two distributions need be known. The Bayes solution requires that the complete distributions be known.

The above results represent the basics of modern credibility theory. There are a number of sources that provide detailed descriptions. Among them are Goovaerts and Hoogstad (1987), Philbrick (1981), and Herzog.

The example in Section B of Chapter 2 is a good illustration of this process. Recall that a single observation had the Poisson distribution with parameter θ_i while the parameter θ had the gamma distribution with parameters α and β. In the notation of this Section, $m(\theta) = \theta$, $m = E(\theta) = \alpha\beta$, $v = Var(\theta) = \alpha\beta^2$, $s(\theta) = \theta$, and $s = E(\theta) = \alpha\beta$. From (5.6) we have

$$Z = \alpha\beta^2/(\alpha\beta^2 + \alpha\beta/t) = \beta t/(\beta t + 1)$$

[13] A second kind of credibility is called limited fluctuation credibility. It originated in the work of Mowbray (1914) and has the drawback that it does not take into account the quality of the data from the other classes. A good modern exposition can be found in Venter (1986).

and so the credibility estimate of the parameter is

$$\frac{\beta t}{\beta t+1}\overline{X}_i+\frac{1}{\beta t+1}\alpha\beta \tag{5.8}$$

which is exactly the same as the Bayes estimate found in Chapter 2. This is one of several cases in which the Bayes estimate and the credibility estimate turn out to be identical.

C. CREDIBILITY ISSUES

The basic model presented above is too simple for practical use. In this section a number of complicating issues are raised and an indication is given of how they have been handled in the past.

The most difficult problem is with the second level parameter μ. In the linear solution presented above all three of the required quantities m, s, and v are functions of μ. Most always there is insufficient knowledge to enable the analyst to specify these values in advance. The only exception is m in the experience rating case. There m is set at the manual premium for the group to which the individual belongs. The point of experience rating is to see what adjustment should be made to m based on the individual's experience. A second way to deal with m is to further restrict the class of estimates. Suppose the only allowable estimates are linear functions of the observations. Further suppose that there are t_i observations from the ith class. Then Bühlmann and Straub (1972) showed that the estimate with the smallest *mse* is

$$Z_i\overline{X}_i+(1-Z_i)\overline{\overline{X}} \quad \text{where} \quad Z_i=\frac{v}{v+s/t_i} \quad \text{and} \quad \overline{\overline{X}}=\frac{\sum Z_j\overline{X}_j}{\sum Z_j}. \tag{5.9}$$

Note that the estimate of m is a credibility weighted average of the class means.

The problem of estimating s and v remains. Bühlmann and Straub (1972) offered a solution to this problem as well. They proposed estimates based on appropriate sums of squares. The sums were then multiplied by an appropriate constant to make them unbiased estimates of the two variances. It turns out that these estimates are the same ones used in the analysis of variance components designs (see Graybill (1961), Chapters 16 and 17, for a discussion of the problem from this viewpoint). The term empirical Bayes has been applied to this procedure. This is the second type of empirical Bayes analysis mentioned in Section E of Chapter 3; the data are used to estimate the parameter (here μ) of the prior distribution. There are at least three problems with this approach. First, as method of moments estimates they are less than optimal. Second, the estimate of v can be negative, a clearly unacceptable value. Finally, when the estimates of v and s are combined to provide an estimate of the credibility factor Z_i the resulting estimate will not be unbiased. An appropriate correction is easy to find when the sample sizes (t_i) are constant and a normal distribution is assumed.

An alternative is the use of pseudo-estimates. These use arbitrarily selected sums of squares with the credibility factors as weights. The estimate involves the quantity being estimated! This requires an iterative solution — an initial value of the parameter is used to compute the sum of squares, yielding an updated value of the parameter, which is then reinserted in the sum of squares, etc. One such estimate was introduced by DeVylder (1981) and another appears in Klugman (1985). Such estimates are usually evaluated by treating the appearance of the parameter in the sum of squares as if it were known. It is then shown that these estimates are unbiased and have a smaller variance then the Bühlmann-Straub estimates. However, this evaluation is artificial, due to the fixing of the parameter in the expression. Another problem with pseudo-estimates is that while the estimate of v cannot be negative, it can be zero. In Klugman (1985) it is demonstrated for a simple model that the pseudo-estimate will be zero in exactly the same situations that the Bühlmann-Straub estimate is negative.

A second problem is obtaining a measure of the quality of the credibility estimate. The *mse* evaluates the quality of the procedure and as such averages over all possible values of $m(\theta)$. We are more interested in the conditional *mse*, fixing θ_i at its true value. For the Bühlmann formula (5.3), we have, from (5.4) and holding all parameters fixed,

$$E_{X\mid\theta}\{[Z\overline{X}_i + (1-Z)m - m(\theta_i)]^2\} = Z^2 s(\theta_i)/t + (1-Z)^2[m(\theta_i) - m]^2. \quad (5.10)$$

This measure of error depends on a number of quantities involving unknown parameters (Z, $s(\theta_i)$, $m(\theta_i)$, and m). Moreover, the quantities Z and m inside the expectation are also random variables since the actual values used will be obtained from the data. None of the standard approaches to credibility analysis provide a method of accounting for this extra variability. The closest attempt in a related field is the parametric empirical Bayes approach advocated by Morris (1983).

A third issue is the extension to more complex models. The extension can take place in a variety of directions. Venter (1985) presents cross-classified and hierarchical models while Hachemeister (1975) presents a model where the pure premiums for a group increase linearly over time. These models propose additional linear relationships among the classes. Another way to add some complexity is to let the pure premiums change in a (structured) stochastic manner over time. These notions have been studied by Meyers (1985) and deJong and Zehnwirth (1983). It is also possible to analyze models with non-linear relationships (DeVylder, 1986). The traditional way to deal with these models is to again find the best linear estimate in terms of *mse*. It will be a function of a number of unknown parameters. The next step is to estimate these parameters by any possible method. Usually this involves the clever manipulation of carefully selected sums of squares. The major drawback to this approach is that each time a new model is proposed the analysis must be created from the beginning. It would be much better to have a method that can be easily adapted to a variety of models.

A fourth issue is the underlying distribution of the observations and the first level parameters. As mentioned earlier, the customary credibility formulas are linear approximations to the Bayes formulas. The quality of the approximation depends on the underlying distributions. For some combinations of model and parameter distributions there is no error at all. For others the error might be considerable. When the quantity to be estimated is the frequency with which claims are filed there is little problem as simple distributions such as the Poisson and negative binomial provide good models. When the amounts of the claims are incorporated there are more problems. The distributions tend to be highly skewed with thick right tails. This is likely to be the situation in which the linear approximation is least likely to do a good job, although little evidence has been provided on this question.

Most all the difficulties mentioned in this section are due to actuaries having spent the past half-century seeking linear solutions to the estimation problem. At a time when the cost of computation was high this was a valuable endeavor. The logical solution is to drop the linear approximation and seek the true Bayesian solution to the problem. It will have the smallest *mse* and provides a variety of other beneficial properties. They are the subject of the next Chapter.

6. THE HIERARCHICAL BAYESIAN APPROACH

A. WHAT IT IS

It is essential at the outset to be clear about what is and what is not a Bayesian approach. In particular, none of the credibility methods being used at this time qualify as true Bayesian analyses. The requirements as introduced in Chapter 2 are few — a prior probability distribution that is determined before the data are collected and a model probability distribution. What we need to do for the credibility problem is identify just where these two items come in.

The key to making this identification is to recognize that both the distribution for the observed losses given the parameter value *and* the distribution of parameter values from class to class are part of the model. We take the view that the distribution of parameter values is not a personal probability or prior opinion, but is the outcome from the unseen process that originally endowed the classes with their propensity to generate claims. Represent the first of these two stages by $f(\boldsymbol{x} \mid \boldsymbol{\theta})$ where $\boldsymbol{x}$ represents all the data and $\boldsymbol{\theta}$ represents the entire collection of parameters needed to determine the distribution of $\boldsymbol{x}$. The nature of this distribution will depend on the classification system being used and the processes that determine both the number and amount of the losses. Returning to the example in Chapter 2, suppose we are observing the number of claims from one randomly selected driver from the collection of all insured drivers. Assume that the number of claims in one year has a Poisson distribution with parameter θ and that the number of claims in different years are independent. If we observe t years of claims from one driver the model is

$$f(\boldsymbol{x} \mid \theta) = e^{-t\theta}\theta^{\Sigma x_i}/\Pi x_i!. \tag{6.1}$$

Most always our objective is to estimate some function of θ. Often it is the mean, $E(X \mid \theta)$. In model (6.1) it is θ. The one unifying feature of all greatest accuracy credibility approaches is the existence of a probability distribution on the parameter θ. Now suppose that we have a model for this distribution — $f(\theta \mid \boldsymbol{\xi})$. In model (6.1) this might be a gamma distribution with parameter $\boldsymbol{\xi} = (\alpha,\beta)'$. With this structure in mind there are three ways for the analyst to proceed.

If $\boldsymbol{\xi}$ is known then the distribution of θ given $\boldsymbol{\xi}$ can be considered as a true prior distribution that expresses our opinion as to the value of θ for a randomly selected driver. The analysis then proceeds as in Chapter 2. The posterior mean was found to be

$$\frac{t\beta}{1+t\beta}\overline{x} + \frac{1}{1+t\beta}\alpha\beta. \tag{6.2}$$

Since α and β were assumed known, the analysis is complete. We can also evaluate the precision of this estimate using the posterior variance

$$Var(\theta \mid \boldsymbol{x}) = (\alpha + t\overline{x})\beta^2/(1+t\beta)^2. \tag{6.3}$$

The problem with this approach is that only rarely will the parameters of the prior distribution be known. The next two methods are ways of dealing with this. The first is called empirical Bayes and is the basis for the Bühlmann-Straub estimate (5.9). The key here is the marginal distribution $m(\boldsymbol{x} \mid \boldsymbol{\xi}) = \int f(\boldsymbol{x} \mid \theta) f(\theta \mid \boldsymbol{\xi}) d\theta$. By considering the sample as coming from this distribution, estimates of $\boldsymbol{\xi}$ can be obtained (usually by the method of moments). These estimates can then be inserted into the Bayes solution $E(\theta \mid \boldsymbol{x},\boldsymbol{\xi})$ to obtain an estimate of θ. A requirement for this or the method to follow is that the data must include observations from a variety of different values of θ. So, this approach is not available for the example given above. There the t observations were all taken from a single value of θ and so could not possibly provide information about the manner in which the parameter is distributed throughout the population. The problems with this empirical Bayes approach were outlined in the previous Chapter.

A better approach is to return to the full Bayesian setting. Since the parameter $\boldsymbol{\xi}$ is unknown, the Bayes approach is to assign a prior distribution $\pi(\boldsymbol{\xi})$ to this parameter. The posterior can now be found as follows[14]

$$f(\boldsymbol{x},\theta,\xi) = f(\boldsymbol{x} \mid \theta)f(\theta \mid \xi)\pi(\xi)$$

$$f(\boldsymbol{x},\theta) = \int f(\boldsymbol{x},\theta,\xi)d\xi$$

$$f(\boldsymbol{x},\xi) = \int f(\boldsymbol{x},\theta,\xi)d\theta$$

$$f(\boldsymbol{x}) = \int f(\boldsymbol{x},\theta)d\theta = \int f(\boldsymbol{x},\xi)d\xi$$

$$\pi^*(\theta \mid \boldsymbol{x}) = \frac{f(\boldsymbol{x},\theta)}{f(\boldsymbol{x})} = \frac{f(\boldsymbol{x} \mid \theta)\int f(\theta \mid \xi)\pi(\xi)d\xi}{\int\int f(\boldsymbol{x} \mid \theta)f(\theta \mid \xi)\pi(\xi)d\xi d\theta}. \tag{6.4}$$

From this density any posterior quantity of interest can be found. It is also possible that information about ξ might be desired. It is available from the posterior density

$$\pi^*(\xi \mid \boldsymbol{x}) = \frac{f(\boldsymbol{x},\xi)}{f(\boldsymbol{x})} = \frac{\pi(\xi)\int f(\boldsymbol{x} \mid \theta)f(\theta \mid \xi)d\theta}{\int\int f(\boldsymbol{x} \mid \theta)f(\theta \mid \xi)\pi(\xi)d\xi d\theta}. \tag{6.5}$$

This has been called a hierarchical Bayes model and was introduced by Lindley and Smith (1972). It is flexible enough to include most any regression, time series, or analysis of variance problem that one is likely to encounter. Generally the experiment itself will suggest the nature of the two higher order distributions, so the only element that may be difficult to specify is the prior distribution on ξ. After that, the only remaining problem is to evaluate the various integrals. Although they appear innocent, in most applications both θ and ξ will be of high dimension and at some point numerical methods will be needed to complete the solution. No clever manipulations will be needed to create variance estimates and confidence intervals. On the other hand, some creativity may reduce the amount of brute force needed to arrive at the answer.

B. AN EXAMPLE

To illustrate these ideas consider again the Poisson model for claim counts. Now suppose we observe each of k different drivers for one year. Driver i has an unknown Poisson parameter θ_i and the Poisson parameters are distributed throughout the population as a gamma distribution with parameters 2 and ξ. Finally, let the prior density on ξ be $f(\xi) \propto 1/\xi$, $\xi > 0$. The elements of the hierarchical model (as set out in (6.4) and (6.5))

[14] The formulas here have θ as a scalar and ξ as a vector to correspond to the Poisson-gamma example. There is no restriction on the dimension of these two quantities.

are as follows:

$$f(x_1,\ldots,x_k \mid \theta_1,\ldots,\theta_k) = \prod_{i=1}^{k} \frac{e^{-\theta_i}\theta_i^{x_i}}{x_i!}$$

$$f(\theta_1,\ldots,\theta_k \mid \xi) = \prod_{i=1}^{k} \theta_i e^{-\theta_i/\xi} \xi^{-2}$$

$$\pi(\xi) \propto \xi^{-1} \tag{6.6}$$

$$\int f(\theta_1,\ldots,\theta_k \mid \xi) f(\xi) d\xi$$

$$= \Big(\prod_{i=1}^{k} \theta_i\Big) \int e^{-\Sigma\theta_i/\xi} \xi^{-2k-1} d\xi = \Big(\prod_{i=1}^{k} \theta_i\Big)\Big(\sum_{i=1}^{k} \theta_i\Big)^{-2k} \tag{6.7}$$

$$\int \cdots \int f(x_i,\ldots,x_k \mid \theta_1,\ldots,\theta_k) f(\theta_1,\ldots,\theta_k \mid \xi) d\theta_1 \cdots d\theta_k$$

$$= \int \cdots \int \prod_{i=1}^{k} \frac{e^{-\theta_i(1+\xi^{-1})}\theta_i^{x_i+1}}{x_i!\xi^2} d\theta_1 \cdots d\theta_k$$

$$= \frac{\prod_{i=1}^{k} \Gamma(x_i+2)}{\xi^{2k}(1+\xi^{-1})^{\Sigma(x_i+2)} \prod_{i=1}^{k} x_i!} \tag{6.8}$$

$$\int \cdots \int \int f(x_i,\ldots,x_k \mid \theta_1,\ldots,\theta_k) f(\theta_1,\ldots,\theta_k \mid \xi) f(\xi) d\xi d\theta_1 \cdots d\theta_k$$

$$= \int \int \cdots \int \xi^{-1} \prod_{i=1}^{k} \frac{e^{-\theta_i(1+\xi^{-1})}\theta_i^{x_i+1}}{x_i!\xi^2} d\theta_1 \cdots d\theta_k d\xi$$

$$= \int \frac{\prod_{i=1}^{k} \Gamma(x_i+2)}{\xi^{2k+1}(1+\xi^{-1})^{\Sigma(x_i+2)} \prod_{i=1}^{k} x_i!} d\xi$$

$$= \left[\prod_{i=1}^{k} (x_i + 1) \right] \int \xi^{k\bar{x} - 1}(1 + \xi)^{-k\bar{x} - 2k} d\xi$$

$$= \left[\prod_{i=1}^{k} (x_i + 1) \right] \frac{\Gamma(k\bar{x})\Gamma(2k)}{\Gamma(k\bar{x} + 2k)} \tag{6.9}$$

$$f(\theta_1, \ldots, \theta_k \mid x_1, \ldots, x_k) \propto \frac{\left(\prod_{i=1}^{k} \theta_i^{x_i + 1} \right) e^{-\Sigma \theta_i} \Gamma(k\bar{x} + 2k)}{\left(\sum_{i=1}^{k} \theta_i \right)^{2k} \left[\prod_{i=1}^{k} (x_i + 1)! \right] \Gamma(k\bar{x})\Gamma(2k)} \tag{6.10}$$

$$f(\xi | x_1, \ldots, x_k) \propto \frac{\Gamma(k\bar{x} + 2k)}{\xi^{2k+1}(1 + \xi^{-1})^{k\bar{x} + 2k}\Gamma(k\bar{x})\Gamma(2k)}$$

$$= \frac{\xi^{k\bar{x} - 1}\Gamma(k\bar{x} + 2k)}{(1 + \xi)^{k\bar{x} + 2k}\Gamma(k\bar{x})\Gamma(2k)}. \tag{6.11}$$

The integral in (6.7) was done by noting that the integrand is a gamma density. The integral in (6.9) was done by making the transformation $\eta = \xi/(1 + \xi)$ and then noting that the result is a beta integral. Even though all constants were carried through in the calculations, there is no guarantee that (6.10) or (6.11) are legitimate probability densities as the prior density $\pi(\xi)$ was not proper. To see that (6.11) integrates to one, use the same change of variable employed to do the integral in (6.9). It does, so the $\propto$ may be replaced by an equals sign. To see if (6.10) integrates to one, make the following change of variables:

$$z = \sum_{i=1}^{k} \theta_i, \;\; y_1 = \theta_1/z, \; \ldots \; , \; y_{k-1} = \theta_{k-1}/z.$$

It is then easy to see that (6.10) would be a density function if the term $\Gamma(2k)$ was removed from the denominator. This makes it easy to get the posterior moments since multiplying the density by θ_s is equivalent to increasing x_s by one (and of course $k\bar{x}$ will also increase by one). Therefore

$$E(\theta_s \mid x_1,\ldots,x_k)$$

$$= \frac{\Gamma(k\overline{x}+2k)}{\Gamma(k\overline{x})\prod_{i=1}^{k}(x_i+1)!} \frac{(x_s+2)\Gamma(k\overline{x}+1)\prod_{i=1}^{k}(x_i+1)!}{\Gamma(k\overline{x}+2k+1)} = \frac{(x_s+2)\overline{x}}{\overline{x}+2}. \tag{6.12}$$

The result is still a linear credibility formula with $Z = \overline{x}/(\overline{x}+2)$.

It is interesting to compare (6.12) to (6.2). The situation is now $t = 1$ since there is just one observation from the sth driver and the credibility factor is $Z = \xi/(1+\xi)$. This implies that ξ is being estimated by $\overline{x}/2$, a not an unreasonable choice since

$$E(\overline{x} \mid \xi) = E[E(\overline{x} \mid \theta_1,\ldots,\theta_k,\xi)] = E(\overline{\theta} \mid \xi) = 2\xi$$

and so the method of moments estimate is indeed $\overline{x}/2$. Instead of this method of moments approach, the hierarchical model allows ξ to be estimated by its posterior mean. From (6.11) it is $k\overline{x}/(2k-1)$, which is slightly larger than the method of moments estimate.

The major advantage of the hierarchical scheme is the ability to evaluate higher moments. In this example

$$E(\theta_s^2 \mid x_1,\ldots,x_k) = \frac{\Gamma(k\overline{x}+2k)}{\Gamma(k\overline{x})\prod_{i=1}^{k}(x_i+1)!} \frac{(x_s+2)(x_s+3)\Gamma(k\overline{x}+2)\prod_{i=1}^{k}(x_i+1)!}{\Gamma(k\overline{x}+2k+2)}$$

$$= \frac{(x_s+2)(x_s+3)(\overline{x})(k\overline{x}+1)}{(\overline{x}+2)(k\overline{x}+2k+1)} \tag{6.13}$$

and so

$$Var_{HB}(\theta_s \mid x_1,\ldots,x_k) = \frac{(x_s+2)(\overline{x})}{(\overline{x}+2)}\left[\frac{(x_s+3)(k\overline{x}+1)}{(k\overline{x}+2k+1)} - \frac{(x_s+2)(\overline{x})}{(\overline{x}+2)}\right]. \tag{6.14}$$

It is interesting to compare this variance formula with (6.3) with $\overline{x}/2$ inserted for ξ. The result is

$$Var_{EB}(\theta_s \mid x_1,\ldots,x_k) = (2+x_s)(\overline{x}/2)^2/(1+\overline{x}/2)^2$$
$$= (2+x_s)\overline{x}^2/(2+\overline{x})^2. \qquad (6.15)$$

The difference is

$$Var_{HB} - Var_{EB} = \frac{2(x_s+2)(x_s+3)\overline{x}}{(k\overline{x}+2k+1)(\overline{x}+2)^2} \qquad (6.16)$$

which is always positive, indicating, as expected, that the empirical Bayes measure of the variance of the estimate understates the actual value. Also note that as $k \rightarrow \infty$ the difference goes to zero. This coincides with the notion that empirical Bayes results are acceptable for large sample sizes.

In fairness to the empirical Bayes approach it should be noted that had the first parameter in the gamma density been left unspecified, the hierarchical Bayes analysis would have been much more difficult. A prior density on α would be needed and integrals with respect to α would have to be evaluated numerically.

C. THE GENERAL HIERARCHICAL MODEL

From the introduction in Section A and the example in Section B it should be clear that the hierarchical approach is able to solve the problems posed in Chapter 5. Regardless of the distribution or the structure of the model, the Bayesian approach yields estimates of all the relevant quantities. In addition, it provides estimates of the quality of these estimates. Finally, estimating future observations and the variance of these estimates is possible. However, this does not mean that implementing the hierarchical Bayesian model is easy. In this Section the general formulas for analyzing the model will be presented. The development follows that in Berger (1985, Chapter 4). The major change is a provision for nuisance parameters at the first level of the hierarchy.

The most general statement of the hierarchical model is (all variables are presented as scalars, but they could be vectors or matrices)

$$f(x \mid \theta, F)$$
$$\pi_1(\theta \mid \mu, G)$$
$$\pi_2(\mu, F, G) \qquad (6.17)$$

The first level indicates how the data (x) depends on the parameter of interest (θ) and some other parameters (F). The second level indicates how the parameter of interest varies throughout the population. The second level distribution depends on some additional unknown parameters (μ and G). Although these two parameters could be listed as one, it will be seen in many of the examples to be presented later that this split will be advantageous. At the third level is a true prior density on the set of nuisance parameters.

The objective is to obtain the posterior distributions in the most efficient manner possible. There are two ways to do this. One is to use the brute force approach as in (6.4) and (6.5). Rewriting those equations for this slightly more general model yields

$$\pi^*(\theta \mid x) = \frac{f(x,\theta)}{m_{22}(x)}$$

$$= \frac{\int\int\int f(x \mid \theta,F)\pi_1(\theta \mid \mu,G)\pi_2(\mu,F,G)d\mu dF dG}{\int\int\int\int f(x \mid \theta,F)\pi_1(\theta \mid \mu,G)\pi_2(\mu,F,G)d\mu dF dG d\theta} \tag{6.18}$$

$$\pi_2^*(\mu,F,G \mid x) = \frac{f(x,\mu,F,G)}{m_{22}(x)}$$

$$= \frac{\pi_2(\mu,F,G)\int f(x \mid \theta,F)\pi_1(\theta \mid \mu,G)d\theta}{\int\int\int\int f(x \mid \theta,F)\pi_1(\theta \mid \mu,G)\pi_2(\mu,F,G)d\mu dF dG d\theta}. \tag{6.19}$$

It is unlikely that these integrals will be easy to do. However, it is often possible to do some of them analytically and the most efficient order is usually the same regardless of the model. Begin by using the law of total probability to write

$$\pi^*(\theta \mid x) = \int\int\int \pi_1^*(\theta \mid x,\mu,F,G)\pi_2^*(\mu,F,G \mid x)d\mu dF dG. \tag{6.20}$$

where the asterisk is used to denote a posterior density. The first density in the integrand is obtained from

$$\pi_1^*(\theta \mid x,\mu,F,G) = \frac{f(x \mid \theta,F)\pi_1(\theta \mid \mu,G)}{m_1(x \mid \mu,F,G)} \tag{6.21}$$

$$m_1(x \mid \mu,F,G) = \int f(x \mid \theta,F)\pi_1(\theta \mid \mu,G)d\theta. \tag{6.22}$$

Often, it will not be necessary to do the integral in (6.22) to obtain the marginal density. If the numerator of (6.21) can be recognized as a standard density (as a function of θ) then the denominator must be whatever needs to be inserted to make the right hand side a legitimate probability density. At the next level

$$\pi_2^*(\mu,F,G \mid x) = \pi_{21}^*(\mu \mid x,F,G)\pi_{22}^*(F,G \mid x) \tag{6.23}$$

$$\pi_{21}^*(\mu \mid x,F,G) = \frac{m_1(x \mid \mu,F,G)\pi_{21}(\mu \mid F,G)}{m_{21}(x \mid F,G)} \tag{6.24}$$

$$m_{21}(x \mid F,G) = \int m_1(x \mid \mu,F,G)\pi_{21}(\mu \mid F,G)d\mu \tag{6.25}$$

$$\pi_{22}^*(F,G \mid x) = \frac{m_{21}(x \mid F,G)\pi_{22}(F,G)}{m_{22}(x)} \tag{6.26}$$

$$\mathrm{m}_{22}(x) = \int \int \mathrm{m}_{21}(x \mid F,G)\pi_{22}(F,G)\mathrm{d}F\mathrm{d}G. \tag{6.27}$$

To use these formulas the third level distribution must be factored as $\pi_2(\mu,F,G) = \pi_{21}(\mu \mid F,G)\pi_{22}(F,G)$. Many times the inner integral in (6.20) can be done analytically. Recognize this by using (6.23) to write

$$\pi_{11}^*(\theta \mid x,F,G) = \int \pi_1^*(\theta \mid x,\mu,F,G)\pi_{21}^*(\mu \mid x,F,G)d\mu \tag{6.28}$$

$$\pi^*(\theta \mid x) = \int \int \pi_{11}^*(\theta \mid x,F,G)\pi_{22}^*(F,G \mid x)dFdG. \tag{6.29}$$

Finally, note that $\pi_{11}^*(\theta \mid x,F,G)$ is the posterior density that would be used in an empirical Bayes analysis.

Of great interest are the posterior mean and variance of the elements of θ. As indicated earlier, the posterior mean is the minimum mean squared error estimate of an element while the variance can be used to create confidence intervals. If θ_i is one such element the direct approach is to use

$$E(\theta_i \mid x) = \int \theta_i \pi^*(\theta \mid x) d\theta \text{ and } E(\theta_i^2 \mid x) = \int \theta_i^2 \pi^*(\theta \mid x) d\theta. \tag{6.30}$$

Each of these expectations is a multiple integral with respect to all the elements of θ and will rarely be amenable to a closed form solution. An alternative approach is to use the following:

$$E(\theta_i \mid x) = E[E(\theta_i \mid x,F,G)] = \int \int m_i(F,G) \pi_{22}^*(F,G \mid x) dF dG \tag{6.31}$$

$$m_i(F,G) = E(\theta_i \mid x,F,G) = \int \theta_i \pi_{11}^*(\theta \mid x,F,G) d\theta. \tag{6.32}$$

It should be noted that all this involves is interchanging the order of integration in (6.30), doing the integral with respect to θ before integrating with respect to F and G. In many cases the integral with respect to θ in (6.32) will be much easier than the one in (6.30). The integral in (6.31) will still have to be done numerically, but will often be of much lower dimension. The variance can be found in much the same manner:

$$\begin{aligned} Var(\theta_i \mid x) &= E[Var(\theta_i \mid x,F,G)] + Var[E(\theta_i \mid x,F,G)] \\ &= \int \int s_i(F,G) \pi_{22}^*(F,G \mid x) dF dG \\ &\quad + \int \int [m_i(F,G)]^2 \pi_{22}^*(F,G \mid x) dF dG - [E(\theta_i \mid x)]^2 \end{aligned} \tag{6.33}$$

$$\begin{aligned} s_i(F,G) &= Var(\theta_i \mid x,F,G) \\ &= \int \theta_i^2 \pi_{11}^*(\theta \mid x,F,G) d\theta - [\int \theta_i \pi_{11}^*(\theta \mid x,F,G) d\theta]^2. \end{aligned} \tag{6.34}$$

These last few formulas indicate the general empirical Bayes approach. (6.32) is used to obtain an estimate that is conditioned upon the nuisance parameters. They are then replaced by estimates obtained by whatever means can be found. The usual empirical Bayes estimate of variability is found by making the same substitution in (6.34), reflecting only the first term in (6.33). Obtaining an estimate of the second term in (6.33) is much more difficult. The parametric empirical Bayes approach of Morris (1983) does this by finding estimates of the variances of the estimates of F and G. This requires the ability not only to find these estimates, but also to find their distribution. For simple situations using sums of squares the estimates will often have a chi-square distribution, but clever approximations will be needed for more complex situations.

In addition to estimating θ we might also be interested in predicting a new observation, say y. The first level distribution is $g(y \mid \theta,F)$ although it may not depend of all the elements of θ and F. The density of this new value is

$$g^*(y \mid x) = \int \int \int f(y \mid \theta,F)\pi_{11}^*(\theta \mid x,F,G)\pi_{22}^*(F,G \mid x)d\theta dFdG. \tag{6.35}$$

Often we will just be interested in the mean and variance of this new value. The mean can be found by inserting y in (6.35) and integrating with respect to y. Doing the integral with respect to y first yields

$$E(y \mid x) = \int \int \int E(y \mid \theta,F)\pi_{11}^*(\theta \mid x,F,G)\pi_{22}^*(F,G \mid x)d\theta dFdG. \tag{6.36}$$

Similarly

$$Var(y \mid x)$$

$$= \int \int \int E(y^2 \mid \theta,F)\pi_{11}^*(\theta \mid x,F,G)\pi_{22}^*(F,G \mid x)d\theta dFdG - E(y \mid x)^2. \tag{6.37}$$

In the applications that are presented in this monograph the integration with respect to θ will be straightforward, leaving integrations similar to those in (6.31) and (6.33).

There is one other posterior distribution of interest. It is of no value in obtaining analytical solutions to the problem, but will be seen to be part of one of the approximation methods discussed in Chapter 7. It is $\pi^*(\theta,F,G \mid x)$ and can be found as follows:

$$\pi^*(\theta,F,G \mid x) = \pi_{11}^*(\theta \mid x,F,G)\pi_{22}^*(F,G \mid x)$$

$$= \frac{f(x \mid \theta,F,G)\pi_{12}^*(\theta \mid F,G)}{m_{21}(x \mid F,G)} \frac{m_{21}(x \mid F,G)\pi_{22}(F,G)}{m_{22}(x)}$$

$$\propto f(x \mid \theta,F)\pi_{12}^*(\theta \mid F,G)\pi_{22}(F,G) \tag{6.38}$$

$$\pi_{12}^*(\theta \mid F,G) = \int \pi_1(\theta \mid \mu,G)\pi_{21}(\mu \mid F,G)d\mu. \tag{6.39}$$

D. SIMPLIFYING ASSUMPTIONS

The general hierarchical model just introduced is much to general to admit a convenient solution. In this Section several restrictions will be placed on the model to make the analysis more tractable. The result is the hierarchical normal linear model; its analysis appears in the next Chapter.

1. Normality

The most significant assumption to be made is that the first two distributions in (6.17) both be multivariate normal. At first glance this seems to be an impractical choice. For frequency counts, a continuous distribution is clearly inappropriate; the Poisson and negative binomial are much more commonly used. For severity or total losses the distribution is seldom symmetric nor does it place probability on negative numbers. Numerous examples of appropriate models are given in Hogg and Klugman (1984).

There are four arguments in support of normality. First, analysis is often done on loss ratios, not the losses themselves. Although this will do nothing to eliminate any skewness in the first level distribution loss ratios may well be symmetrically distributed at the second level. That is, the class to class deviations are just as likely to be large as they are to be small. The second justification is that the normal distribution is easy to work with. We will see in the *next* Chapter that formulas can be derived for the various posterior quantities of interest. This will enable us to take advantage of the power of a true Bayesian analysis. As our abilities to do numerical integration increase this will be less of a necessity. The third argument also has to do with convenience. It may be the case that our model includes dependent observations. The multivariate normal model is one of the few that easily allows for dependence. The fourth argument is the most compelling. It has been known for some time that the Bayes solution (5.2) and the credibility (linear Bayes) solution (5.3) will be identical whenever the first level distribution is a member of the linear exponential family and the second level distribution is the natural conjugate prior (Ericson, 1969, and Jewell, 1974). There has been speculation (Goel, 1982) that these are the only distributions that possess this property. To a certain extent, those who are willing to accept a linear solution should be equally comfortable restricting attention to models from the linear exponential family. While there are pairs other than normal-normal that fit this description (e.g., Poisson-gamma, binomial-beta), it is the only one for which the optimal variance estimates are sums of squares. This is the approach taken by empirical Bayes analysts. So, using the normal distribution should lead us to roughly the same estimate as those who use the first two moments. The normal model brings two advantages. As mentioned before, with a complete distributional model we can do a full Bayesian analysis and thereby obtain other quantities such as prediction intervals. Also, we are displaying our assumptions up front, and know what needs to be done if they are to be changed.

When analyzing non-normal observations the conventional recommendation is to look for a transformation that will make the data more normal. This technique may also be of assistance here. For example, suppose for a given exposure P the number of claims X has a Poisson distribution with parameter $P\lambda$. The relative frequency X/P has mean λ, variance λ/P, and skewness $1/\sqrt{\lambda P}$. Now let $Y = 2\sqrt{X/P}$. Asymptotically (as λ gets large) Y has mean $2\sqrt{\lambda}$ and variance $1/P$. A bonus from this transformation is that the variance no longer depends on the parameter of interest, λ. In fact, it does not depend on any unknown parameter! In Table 6.1 values for six levels of $P\lambda$ are given. At a value of 10 the approximation begins to look very good. The major drawback of using a transformation is that the inverse transformation must be used to get an answer that is useful. An example using this transformation will be presented in Chapter 9.

Table 6.1

Square Root Transformation of a Poisson Variable

$P\lambda$	1	2	5	10	20	50
$PE(X/P)$	1	2	5	10	20	50
$P^2Var(X/P)$	1	2	5	10	20	50
$Skew(X/P)$	1	0.707	0.447	0.316	0.224	0.141
$\sqrt{P}E(\sqrt{X/P})$	1.546	2.538	4.342	6.241	8.887	14.106
$PVar(2\sqrt{X/P})$	1.609	1.560	1.144	1.045	1.020	1.008
$Skew(2\sqrt{X/P})$	−0.056	−0.414	−0.260	−0.101	−0.062	−0.037

The above does not mean that non-normal models should be ignored and in Chapter 9 we drop the assumption of normality.

2. Linearity

As the normal distribution is characterized by its mean and variance, any further simplifications will relate to the specification of these quantities. Restriction to linear specifications means that the population means must be linear functions of the unknown parameters. For example, a specification that the means are increasing arithmetically over time ($\mu_t = a + bt$) is acceptable as the mean is a linear function of the unknown parameters a and b, while geometrically increasing means ($\mu_t = a(1+b)^t$) is not. As will be seen in the next Chapter the commonly used models happen to be linear.

This is another case where transformations may be useful. With geometrically increasing means, the natural logarithm ($ln\mu_t = lna + tln(1+b)$) is linear in the parameters lna and $ln(1+b)$. While taking logarithms of the observations does not automatically change the mean to its logarithm, the error may be immaterial. Of more concern is the effect taking the logarithm has on the variance. This will be explored in the next Section.

3. Variance Independent of the Mean

The goal of the investigation is to estimate the mean. This will be much easier if the variance (and covariances) does not depend on the mean. For the normal distribution, this is usually the case, but is almost certainly not the case for other distributions. For example, with the Poisson distribution the variance is equal to the mean and so an assumption of constant variance is unlikely to be reasonable. This is the advantage of the square root transformation mentioned in Subsection 1.

On the other hand, transformations can also create problems. Suppose we believe that the logarithm of the observations has a normal distribution with a mean that is arithmetically increasing over time. Further suppose that the variances of the transformed observations are inversely proportional to the exposure. That is, if X_t is the original observation at time t then $lnX_t \sim N(a+bt, \sigma^2/P_t)$ and X_t has mean μ_t and variance s_t^2. From the lognormal distribution we then have

$$\mu_t = exp(a + bt + \sigma^2/2P_t) \text{ and}$$

$$s_t^2 = exp(2a + 2bt + \sigma^2/P_t)[exp(\sigma^2/P_t) - 1] \doteq \mu_t^2\sigma^2/P_t.$$

Assuming that transformed variables have a convenient representation implies that the actual observations may have a variance pattern that we do not believe. Similarly, if we assign a realistic value to s_t^2 then the variance of the transformed variable will depend on a and b.

7. THE HIERARCHICAL NORMAL LINEAR MODEL

A. THE MODEL

In this Chapter one more restriction to the normal model of Chapter 6 will be imposed: linearity in the parameters. Within this model most all standard situations involving severity, pure premiums, or loss ratios can be handled. The only reasonable case that cannot be handled is the Poisson model for frequency. This will be covered in Chapter 9.

The model itself is easily written in the three part framework of (6.17). The three levels are

$$\begin{aligned} &\boldsymbol{x} \mid \boldsymbol{\theta}, F \sim N(A\boldsymbol{\theta}, F) \\ &\boldsymbol{\theta} \mid \boldsymbol{\mu}, G \sim N(B\boldsymbol{\mu}, G) \\ &\pi_2(\boldsymbol{\mu}, F, G) \end{aligned} \tag{7.1}$$

where $\boldsymbol{x}$ is an $N \times 1$ vector containing the data, $\boldsymbol{\theta}$ is a $k \times 1$ vector containing the parameters of interest, $\boldsymbol{\mu}$ is a $z \times 1$ vector of hyperparameters, A $(N \times k)$ and B $(k \times z)$ are known matrices, and F $(N \times N)$ and G $(k \times k)$ are unknown positive definite covariance matrices. The two covariance matrices may have some additional structure. The third level distribution will be left unspecified for now. It will be discussed in Section D of this Chapter. The model in (7.1) is the form of the original hierarchical normal linear model introduced by Lindley and Smith (1972). It has been used in a variety of situations. Examples can be found in Miller and Fortney (1984), Racine, et. al. (1986) and Smith and West (1983).

B. EXAMPLES – DESCRIPTION

In this section a number of special cases of (7.1) will be presented. These will cover most of the commonly used models for estimation of severity, pure premiums, or loss ratios.

1. One-way

The simplest case corresponds to the one analyzed by Bühlmann and Straub (1972). It is also the model studied by Meyers (1984). The first level consists of n_i observations from each of k groups. Let x_{ij} be the jth observation from the ith group. Then

$$x_{ij} \mid \theta_i, \sigma^2 \sim N(\theta_i, \sigma^2/P_{ij}) \quad i=1,\ldots,k \quad j=1,\ldots,n_i \tag{7.2}$$

and the $N = \Sigma n_i$ observations are conditionally independent (given $\theta_1,\ldots,\theta_k$ and σ^2). The P_{ij} are known values that are proportional to the exposure that produced the observation. The assumption of a constant variance, σ^2, is done for simplification and it should be noted that even though the model introduced by Bühlmann and Straub allowed for a different variance for each group, their answer assigns the same value to each group. To place (7.2) in the notation of (7.1) let the vector $\boldsymbol{x}$ be

$$\boldsymbol{x}' = (x_{11},\ldots,x_{1n_1},x_{21},\ldots,x_{kn_k}) \tag{7.3}$$

and the matrix A be

$$\begin{bmatrix} \mathbf{1}_{n_1} & \mathbf{0}_{n_1} & \cdots & \mathbf{0}_{n_1} \\ \mathbf{0}_{n_2} & \mathbf{1}_{n_2} & \cdots & \mathbf{0}_{n_2} \\ \vdots & \vdots & & \vdots \\ \mathbf{0}_{n_k} & \mathbf{0}_{n_k} & \cdots & \mathbf{1}_{n_k} \end{bmatrix} \tag{7.4}$$

where $\mathbf{1}_m$ is an $m \times 1$ vector of ones and $\mathbf{0}_m$ is an $m \times 1$ vector of zeros. The covariance F is diagonal with $\sigma^2/P_{11},\ldots,\sigma^2/P_{kn_k}$ on the diagonal.

The second level is

$$\theta_i \mid \mu, \tau^2 \sim N(\mu, \tau^2) \quad i=1,\ldots,k \tag{7.5}$$

where $\theta_1,\ldots,\theta_k$ are independent given μ and τ^2. To place (7.5) in the format of (7.1) let $B = \mathbf{1}_k$, $\boldsymbol{\mu} = \mu$, a scalar, and $F = \tau^2 I_k$, where I_k is the $k \times k$ identity matrix. The rationale for this model is that each of the k classes has been given a propensity to produce losses, θ_i. The source of this parameter is irrelevant but the fact that they are independent and identically distributed means that prior to collecting any data there is no reason to believe that one particular group has a higher parameter value than any other group or that if you were told the parameter value for one group, the probability density for the parameter for some other group would change. A way to visualize the first statement is to place someone in a position where they knew the definition of each of the k groups and had a list of the k parameter values. The person is then asked to try to match the parameter values with the groups. The notion of identical distributions means that the person would have only the random chance (1 in $k!$) of getting them all correct. Note that this is not likely to be true when the observations are actual losses. For many types of insurance we would have a good prior knowledge as to which groups have the high parameter values and which have the low ones. That is one reason (as noted in Chapter 6) that loss ratios are more appropriate than the losses themselves. The use of independence can also be understood by looking at this matching situation. Now suppose you are told one of the group-parameter pairs. Would this improve your ability to match the others? This is probably more realistic than the assumption of identical distributions. It should be further noted that there is nothing wrong with using this model for actual losses. The only drawback is that we are doing the analysis by ignoring relevant prior information, something a good Bayesian analyst should avoid. On the other hand, incorporating prior orderings is very difficult. To see how much extra work this can involve, see Broffitt (1984).

To complete this simple model, note that the third level distribution is $\pi_2(\mu,\sigma^2,\tau^2)$, a three-dimensional distribution. As will be seen later, all the formulas for this model can be either done analytically or reduced to a one-dimensional numerical integral. For future reference the model just discussed will be called the one-way model.

2. Two-way

There are several directions in which this simple model can be generalized. One way is to add more structure at the first or second level. A natural extension is to introduce a two-way classification. For example, in automobile insurance, the insureds could be classified by both territory and personal characteristics (various combinations of age, sex, marital status, etc.). The observations will be X_{ijt}, the loss (or loss ratio) in year t

from insureds in territory i and classification group j. To set up the model we need to know what the answers are to look like. One way would be for each of the classes to have its own parameter, θ_{ij}. The first level is

$$x_{ijt} \mid \theta_{ij}, \sigma^2 \sim N(\theta_{ij}, \sigma^2/P_{ijt}). \tag{7.6}$$

If we postulate an additive model for the relationship of the two dimensions the second level would be expressed in two steps:

$$\theta_{ij} \mid \alpha_i, \beta_j, \gamma^2 \sim N(\alpha_i + \beta_j, \gamma^2) \tag{7.7}$$

$$\alpha_i \mid \tau_\alpha^2 \sim N(\mu/2, \tau_\alpha^2) \quad \beta_j \mid \tau_\beta^2 \sim N(\mu/2, \tau_\beta^2) \tag{7.8}$$

with all of these random variables conditionally independent. Letting both the α's and the β's have mean $\mu/2$ is not a simplification since the only relevant quantity is the sum of the two means. Recalling that with the hierarchical model the ultimate goal is to estimate the first level parameters, we see that this formulation does indeed produce individual estimates for each class. That also means the final estimates will not fit an additive model. This is due to (7.7) which allows each class mean to vary from the strict additive model. The amount of deviation allowed will depend on the amount of exposure in that class and the magnitude of γ^2. To create the three level hierarchical model (7.7) and (7.8) will have to be combined into one statement concerning the conditional distribution

$$\theta_{ij} \mid \mu, \gamma^2, \tau_\alpha^2, \tau_\beta^2.$$

Since all the distributions involved are normal, this one will also be normal. The parameters are

$$E(\theta_{ij} \mid \mu, \gamma^2, \tau_\alpha^2, \tau_\beta^2) = E[E(\theta_{ij} \mid \mu, \alpha_i, \beta_j, \gamma^2, \tau_\alpha^2, \tau_\beta^2)]$$
$$= E(\alpha_i + \beta_j \mid \mu, \gamma^2, \tau_\alpha^2, \tau_\beta^2) = \mu$$

$$Var(\theta_{ij} \mid \mu, \gamma^2, \tau_\alpha^2, \tau_\beta^2) = E[Var(\theta_{ij} \mid \mu, \alpha_i, \beta_j, \gamma^2, \tau_\alpha^2, \tau_\beta^2)]$$
$$+ Var[E(\theta_{ij} \mid \mu, \alpha_i, \beta_j, \gamma^2, \tau_\alpha^2, \tau_\beta^2)]$$

$$= E(\gamma^2 \mid \mu,\gamma^2,\tau_\alpha^2,\tau_\beta^2) + Var(\alpha_i + \beta_j \mid \mu,\gamma^2,\tau_\alpha^2,\tau_\beta^2)$$

$$= \gamma^2 + \tau_\alpha^2 + \tau_\beta^2$$

$$Cov(\theta_{ij},\theta_{ij'} \mid \mu,\gamma^2,\tau_\alpha^2,\tau_\beta^2) = E[Cov(\theta_{ij},\theta_{ij'} \mid \mu,\alpha_i,\beta_j,\gamma^2,\tau_\alpha^2,\tau_\beta^2)]$$

$$+ Cov[E(\theta_{ij},\theta_{ij'} \mid \mu,\alpha_i,\beta_j,\gamma^2,\tau_\alpha^2,\tau_\beta^2)]$$

$$= \quad E(0 \mid \mu,\gamma^2,\tau_\alpha^2,\tau_\beta^2) + Cov(\alpha_i + \beta_j,\alpha_i + \beta_{j'} \mid \mu,\gamma^2,\tau_\alpha^2,\tau_\beta^2)$$

$$= \tau_\alpha^2, \quad j \neq j'$$

$$Cov(\theta_{ij},\theta_{i'j} \mid \mu,\gamma^2,\tau_\alpha^2,\tau_\beta^2) = \tau_\beta^2, \quad i \neq i'$$

$$Cov(\theta_{ij},\theta_{i'j'} \mid \mu,\gamma^2,\tau_\alpha^2,\tau_\beta^2) = 0, \quad i \neq i', \;\; j \neq j'. \tag{7.9}$$

The final level requires a prior distribution on the nuisance parameters, $\pi_2(\mu,\gamma^2,\tau_\alpha^2,\tau_\beta^2)$. Note that the additive structure now appears in the second level covariance.

The other way to approach this two-way model is to insist that the final estimates be additive. This is equivalent to setting $\gamma^2 = 0$ and then combining (7.6) and (7.7). The two levels become:

$$x_{ijt} \mid \alpha_i,\beta_j \sim N(\alpha_i + \beta_j,\sigma^2/P_{ijt})$$

$$\alpha_i \mid \tau_\alpha^2 \sim N(\mu/2,\tau_\alpha^2) \quad \beta_j \mid \tau_\beta^2 \sim N(\mu/2,\tau_\beta^2). \tag{7.10}$$

This is much simpler, but does not allow for departures from the linear structure for classes with a large exposure.

3. Linear Trend

The next model was introduced by Hachemeister (1975) and will be called the linear trend model. It is an attempt to incorporate trend factors into the model. In most applications trend effects are eliminated before the credibility formulas are applied. This is generally done by applying inflation factors to the past losses. The Hachemeister model lets the credibility procedure estimate the trend factors at the same time the class means are being estimated. This model is a one-way model where x_{ij} is the loss from group i in year j. The two levels are

$$x_{ij} \mid \alpha_i, \beta_i, \sigma^2 \sim N(\alpha_i + \beta_i j, \sigma^2 / P_{ij})$$

$$\alpha_i \mid \mu_\alpha, \tau_\alpha^2 \sim N(\mu_\alpha, \tau_\alpha^2), \quad \beta_i \mid \mu_\beta, \tau_\beta^2 \sim N(\mu_\beta, \tau_\beta^2)$$

$$Cov(\alpha_i, \beta_i \mid \tau_{\alpha\beta}) = \tau_{\alpha\beta}. \tag{7.11}$$

This model allows for a different straight line to be used for each class. If the inflation rate is assumed to be the same for all classes only one β value is needed. Also note that this model uses a linear trend. Over the short period (3 to 7 years) that is usually used for rate making this is not an unreasonable approximation to the more commonly used exponential trend. To incorporate exponential trend the first level mean would be $\alpha_i(1+\beta_i)^j$ which does not fit the linearity restriction of this Chapter.

4. Kalman Filter

We now return to the Kalman filter model that was introduced in Chapter 4. As in that Chapter, the objective is to reflect stochastically changing parameters. Only two cases will be considered here in detail. The first is the autoregressive process for the class mean. For this to make sense we must be in the situation where the index j represents the time at which the observation was taken. The first two levels of the hierarchical model are just the two equations of the Kalman filter model

$$x_{ij} \mid \theta_{ij}, \sigma^2 \sim N(\theta_{ij}, \sigma^2 / P_{ij})$$

$$\theta_{ij} \mid \theta_{i,\, j-1}, \rho, \eta_i, \gamma^2 \sim N(\rho\theta_{i,\, j-1} + (1-\rho)\eta_i, \gamma^2). \tag{7.12}$$

In the notation of Chapter 4 we have

$$A = [1 \;\; 0], \quad \boldsymbol{\mu}_j = [\theta_{ij} \;\; \eta_i]', \quad F = \sigma^2 / P_{ij},$$

$$B = \begin{bmatrix} \rho & 1-\rho \\ 0 & 1 \end{bmatrix}, \quad G = \begin{bmatrix} \gamma^2 & 0 \\ 0 & 0 \end{bmatrix}. \tag{7.13}$$

An interesting feature of this model is the unconditional (with respect to θ_{ij}) variance of an observation. That is,

$$\begin{aligned} &Var(x_{ij} \mid \eta_i, \rho, \sigma^2, \gamma^2) \\ &\quad = Var[E(x_{ij} \mid \theta_{ij}, \eta_i, \rho, \sigma^2, \gamma^2)] + E[Var(x_{ij} \mid \theta_{ij}, \eta_i, \rho, \sigma^2, \gamma^2)] \\ &\quad = Var(\theta_{ij} \mid \eta_i, \rho, \sigma^2, \gamma^2) + E(\gamma^2 \mid \eta_i, \rho, \sigma^2, \gamma^2) \\ &\quad = \sigma^2/P_{ij} + \gamma^2. \end{aligned}$$

There is some empirical evidence that such a model is a valid description of the relationship between the pure premium and the size of the risk (Hewitt, 1967, Meyers and Schenker, 1983, and Meyers, 1985).[15] A key consequence of this model is that as P_{ij} goes to infinity, the credibility does not go to one. In the one-way model this is expected because an infinite amount of data allows perfect estimation of the unchanging parameter and thus there is no need to bring in information from other classes. In this model the parameter is changing over time and therefore the ability to perfectly estimate the past does not imply perfect predictions of the future.

There are two special cases of interest. If $\rho = 0$, each year's parameter can be considered as a random observation from a distribution with an unchanging mean. If $\rho = 1$, each year's parameter is the previous year's parameter plus a random adjustment. When ρ is between these values we can think of a particular year's parameter as one that reflects the make-up of those individuals who are in the insured group that year. A proportion of them, ρ, continue in the *next* year (and have an expected value equal to last year's value), while the remainder, $1-\rho$, are drawn from the general population (and have an expected value equal to the unchanging class mean).

What differentiates this formulation from that in Chapter 4 is that there are now k different Kalman filter processes operating at once, one for

[15] These authors were dealing with the case where P_{ij} is the premium and so X_{ij} is the loss ratio. It is often the case that the premium is proportional to the exposure and so their observation would still apply when the quantities are exposure and pure premium respectively.

each of the classes. To complete the hierarchical model we must describe the class-to-class relationship of the parameters. There are two ways to do this. The first is to establish the relationship at time zero. This would be a multivariate normal distribution on the parameters θ_{i0} and η_i. The final level would then be a prior distribution on the nuisance parameters (σ^2, γ^2, ρ, and the mean and variance of the multivariate normal distribution just introduced).

The second model is an alternative to the linear trend model. In each period the mean is incremented by a constant amount plus some noise. The equations and matrices are:

$$x_{ij} \mid \theta_{ij}, \sigma^2 \sim N(\theta_{ij}, \sigma^2/P_{ij})$$

$$\theta_{ij} \mid \theta_{i,\,j-1}, \eta_{i,\,j-1}, \gamma_1^2, \gamma_2^2 \sim N(\theta_{i,\,j-1} + \eta_{i,\,j-1}, \gamma_1^2)$$

$$\eta_{ij} \mid \theta_{i,\,j-1}, \eta_{i,\,j-1}, \gamma_1^2, \gamma_2^2 \sim N(\eta_{i,\,j-1}, \gamma_2^2) \tag{7.14}$$

and

$$A = [1 \quad 0], \quad \boldsymbol{\mu}_j = [\theta_{ij}, \ \eta_{ij}]', \quad F = \sigma^2/P_{ij},$$

$$B = \begin{bmatrix} 1 & 1 \\ 0 & 1 \end{bmatrix}, \quad G = \begin{bmatrix} \gamma_1^2 & 0 \\ 0 & \gamma_2^2 \end{bmatrix}. \tag{7.15}$$

The third equation in (7.14) allows the slope to change slightly over time.

The only tractable solution for these formulations is to find the maximum likelihood estimate of the these parameters and then insert them in the Kalman filter equations. This is equivalent to an empirical Bayes approach and will fail to capture the error involved in estimating the nuisance parameters. A second method is to run the Kalman filter models separately on each class. The result will be a time n distribution for the parameters in each class. At this point a second level distribution can be created that describes the class-to-class variation. The third level is a non-informative prior distribution on the parameters of the second level distribution. A detailed explanation of this process is given in Ledolter,

Klugman, and Lee (1991). An example is given in the next Chapter.

5. Graduation

In Chapter 4 a method for graduating mortality tables was introduced. At that time the second level distribution was assumed known. When this method is applied in practice the single parameter of this distribution must be selected by the analyst. This is usually done by trial and error. By casting this problem in the *HNLM* framework we should be able to let the data select the parameter.

As before, let $\boldsymbol{x}$ be the vector of observations and $\boldsymbol{\theta}$ be the vector of true values. The first level is then

$$\boldsymbol{x} \sim N(\boldsymbol{\theta}, F) \tag{7.16}$$

where F is a known diagonal vector with ith element $x_i(1-x_i)/n_i$ where n_i is the size of the sample used to obtain the estimate x_i. The matrix A is thus the identity matrix.

The second level is

$$\boldsymbol{\theta} \sim N(\mathbf{0}, \tau^2(K'K)^{-1}) \tag{7.17}$$

where $\mathbf{0}$ is a vector of 0's and K is the matrix described in (4.3) with additional rows inserted at the beginning. This will be described in detail in the example given in Chapter 8. The matrix B is the identity matrix, and unlike most cases of the *HNLM* the vector $\boldsymbol{\mu}$ is known.

The final level is the prior distribution is

$$\pi_2(\tau^2). \tag{7.18}$$

C. ANALYSIS OF THE MODEL

1. Preliminaries

We return to the general hierarchical normal linear model as set forth in (7.1). The analysis will proceed as outlined in (6.17) through

(6.29). The basic tool required is the matrix formula for completing the square. The first fact is that

$$\int exp[-(\boldsymbol{x}-\boldsymbol{\mu})'A(\boldsymbol{x}-\boldsymbol{\mu})/2]d\boldsymbol{x} = |A|^{-1/2}(2\pi)^{p/2} \tag{7.19}$$

where $\boldsymbol{x}$ is a $p \times 1$ vector. This is just the integral for finding the constants in a multivariate normal distribution. The second fact is

$$\boldsymbol{x}'A\boldsymbol{x} - 2\boldsymbol{x}'B\boldsymbol{g} = (\boldsymbol{x} - A^{-1}B\boldsymbol{g})'A(\boldsymbol{x} - A^{-1}B\boldsymbol{g}) - \boldsymbol{g}'BA^{-1}B\boldsymbol{g} \tag{7.20}$$

provided A is a nonsingular matrix. These two results can be combined to produce

$$\int exp[-(\boldsymbol{x}'A\boldsymbol{x} - 2\boldsymbol{x}'B\boldsymbol{g})/2]\mathrm{d}\boldsymbol{x} = (2\pi)^{p/2}|A|^{-1/2}exp(\boldsymbol{g}'B'A^{-1}B\boldsymbol{g}/2). \tag{7.21}$$

Finally, whenever an expression is encountered that looks like the integrand in (7.21) and the variable $\boldsymbol{x}$ appears nowhere else, then as a function of $\boldsymbol{x}$ it is proportional to a multivariate normal density with mean $A^{-1}B\boldsymbol{g}$ and covariance A^{-1}.

Although many of the intermediate densities will turn out to have normal distributions, the two densities of greatest interest, $\pi^*(\boldsymbol{\theta} \mid \boldsymbol{x})$ and $\pi^*(\boldsymbol{\mu} \mid \boldsymbol{x})$, will not. There are two ways to approximate these distributions. The first is the Bayesian Central Limit Theorem. A version of this Theorem is given in Berger (1985, page 224). It states that if $\boldsymbol{x}$ consists of independently and identically distributed random variables each depending on the parameter $\boldsymbol{\theta}$ and if both the model and the prior densities are twice differentiable with respect to $\boldsymbol{\theta}$ near the maximum likelihood estimate then the posterior density of $\boldsymbol{\theta}$ given $\boldsymbol{x}$ is approximately multivariate normal with the corresponding posterior mean vector and covariance. In addition, if the posterior mean and covariance are not known they can be approximated by the posterior mode and inverse of the matrix of second derivatives of the log posterior, evaluated at the mode. The normal approximation is still valid in this second case although it is not as accurate. These are the same approximations as in Chapter 3.

2. The Analysis

Begin the analysis by performing a simplification of the level one density. It is

$$f(\boldsymbol{x} \mid \boldsymbol{\theta},F) \propto |F|^{-1/2} exp[-(\boldsymbol{x} - A\boldsymbol{\theta})'F^{-1}(\boldsymbol{x} - A\boldsymbol{\theta})/2]$$

$$= |F|^{-1/2} exp\{-[\boldsymbol{x}'F^{-1}\boldsymbol{x} - 2\boldsymbol{\theta}'A'F^{-1}\boldsymbol{x} + \boldsymbol{\theta}'A'F^{-1}A\boldsymbol{\theta}]/2\}.$$

Complete the square with respect to $\boldsymbol{\theta}$ and let $\Sigma^{-1} = A'F^{-1}A$ to obtain

$$f(\boldsymbol{x} \mid \boldsymbol{\theta},F) \propto |F|^{-1/2} exp[-(\boldsymbol{\theta} - \Sigma A'F^{-1}\boldsymbol{x})'\Sigma^{-1}(\boldsymbol{\theta} - \Sigma A'F^{-1}\boldsymbol{x})/2]$$

$$\times\ exp[-(\boldsymbol{x}'F^{-1}\boldsymbol{x} - \boldsymbol{x}'F^{-1}A\Sigma A'F^{-1}\boldsymbol{x})/2]. \qquad (7.22)$$

Now let $\hat{\boldsymbol{\theta}} = \Sigma A'F^{-1}\boldsymbol{x}$, the least squares estimate of $\boldsymbol{\theta}$. The second exponential term can be written $S = (\boldsymbol{x} - A\hat{\boldsymbol{\theta}})'F^{-1}(\boldsymbol{x} - A\hat{\boldsymbol{\theta}})$. Note that the only unknown quantity in S and Σ is the covariance F. Multiplying and dividing by $|\Sigma|^{1/2}$ yields

$$f(\boldsymbol{x} \mid \boldsymbol{\theta},F) \propto |\Sigma|^{-1/2} exp[-(\boldsymbol{\theta} - \hat{\boldsymbol{\theta}})'\Sigma^{-1}(\boldsymbol{\theta} - \hat{\boldsymbol{\theta}})/2]$$

$$\times |\Sigma|^{1/2} |F|^{-1/2} exp(-S/2)$$

$$= f_1(\hat{\boldsymbol{\theta}} \mid \boldsymbol{\theta},\Sigma) f_2(S \mid F) \qquad (7.23)$$

where f_1 is a normal density with mean $\boldsymbol{\theta}$ and covariance Σ.

The next step is to obtain $\pi_1^*(\boldsymbol{\theta} \mid \boldsymbol{x},\boldsymbol{\mu},F,G)$ using the numerator of (6.21). As a function of $\boldsymbol{\theta}$ it is (using (7.23))

$$\pi_1^*(\boldsymbol{\theta} \mid \boldsymbol{x},\boldsymbol{\mu},F,G) \propto f(\boldsymbol{x} \mid \boldsymbol{\theta},F)\pi_1(\boldsymbol{\theta} \mid \boldsymbol{\mu},G) \propto f_1(\hat{\boldsymbol{\theta}} \mid \boldsymbol{\theta},\Sigma)\pi_1(\boldsymbol{\theta} \mid \boldsymbol{\mu},G)$$

$$\propto exp[-(\hat{\boldsymbol{\theta}} - \boldsymbol{\theta})'\Sigma^{-1}(\hat{\boldsymbol{\theta}} - \boldsymbol{\theta})/2 - (\boldsymbol{\theta} - \mathrm{B}\boldsymbol{\mu})'G^{-1}(\boldsymbol{\theta} - B\boldsymbol{\mu})/2]$$

$$\propto exp[-\boldsymbol{\theta}'(\Sigma^{-1} + G^{-1})\boldsymbol{\theta}/2 + 2\boldsymbol{\theta}'(\Sigma^{-1}\hat{\boldsymbol{\theta}} + G^{-1}B\boldsymbol{\mu})/2]. \qquad (7.24)$$

and so

$$\boldsymbol{\theta} \mid \boldsymbol{x},\boldsymbol{\mu},F,G \sim N[(\Sigma^{-1} + G^{-1})^{-1}(\Sigma^{-1}\hat{\boldsymbol{\theta}} + G^{-1}B\boldsymbol{\mu}),\ (\Sigma^{-1} + G^{-1})^{-1}]. \qquad (7.25)$$

To avoid excessive inversions let $W = (\Sigma + G)^{-1}$ and then write the covariance and mean as

$$V_1 = (\Sigma^{-1} + G^{-1})^{-1} = \Sigma - \Sigma W \Sigma \tag{7.26}$$

$$\begin{aligned}\boldsymbol{\mu}_1 &= (\Sigma^{-1} + G^{-1})^{-1}(\Sigma^{-1}\hat{\boldsymbol{\theta}} + G^{-1}B\boldsymbol{\mu}) \\ &= (\Sigma - \Sigma W \Sigma)\Sigma^{-1}\hat{\boldsymbol{\theta}} + \Sigma(G+\Sigma)^{-1}GG^{-1}B\boldsymbol{\mu} \\ &= (I - \Sigma W)\hat{\boldsymbol{\theta}} + \Sigma W B\boldsymbol{\mu}. \end{aligned}\tag{7.27}$$

This is the multivariate equivalent of the Bühlmann solution (5.3). As noted in Chapter 5, this is one of the situations in which the Bayes and Bühlmann solutions will be the same.

The next step is to get the marginal distribution $m_1(\boldsymbol{x} \mid \boldsymbol{\mu},F,G)$ from (6.22). It is the convolution of two normal distributions and so will itself be normal. The relevant parameters are

$$E(\boldsymbol{x} \mid \boldsymbol{\mu},F,G) = E[E(\boldsymbol{x} \mid \boldsymbol{\theta},\boldsymbol{\mu},F,G)] = E(A\boldsymbol{\theta} \mid \boldsymbol{\mu},F,G) = AB\boldsymbol{\mu}$$

$$\begin{aligned}Var(\boldsymbol{x} \mid \boldsymbol{\mu},F,G) &= Var[E(\boldsymbol{x} \mid \boldsymbol{\theta},\boldsymbol{\mu},F,G)] + E[Var(\boldsymbol{x} \mid \boldsymbol{\theta},\boldsymbol{\mu},F,G)] \\ &= Var(A\boldsymbol{\theta} \mid \boldsymbol{\mu},F,G) + E(F \mid \boldsymbol{\mu},F,G) \\ &= AGA' + F. \end{aligned}\tag{7.28}$$

Also worth noting is that this marginal density can be factored in the same way that $f(\boldsymbol{x} \mid \boldsymbol{\theta},F)$ was factored to give

$$m_1(\boldsymbol{x} \mid \boldsymbol{\mu},F,G) = f_2(S \mid F)m_{11}(\hat{\boldsymbol{\theta}} \mid \boldsymbol{\mu},F,G) \tag{7.29}$$

$$\hat{\boldsymbol{\theta}} \mid \boldsymbol{\mu},F,G \sim N(B\boldsymbol{\mu},W^{-1}). \tag{7.30}$$

This follows from the fact that S and $\hat{\boldsymbol{\theta}}$ are conditionally independent given F (see (7.23)) and then arguing as in (7.28) to get the parameters of the normal distribution.

Next obtain $\pi^*_{21}(\boldsymbol{\mu} \mid \boldsymbol{x},F,G)$ using (6.24). As a function of $\boldsymbol{\mu}$ it is (using (7.29))

$$\pi_{21}^*(\boldsymbol{\mu} \mid \boldsymbol{x},F,G) \propto m_1(\boldsymbol{x} \mid \boldsymbol{\mu},F,G)\pi_{21}(\boldsymbol{\mu} \mid F,G)$$

$$\propto m_{11}(\hat{\boldsymbol{\theta}} \mid \boldsymbol{\mu},F,G)\pi_{21}(\boldsymbol{\mu} \mid F,G). \tag{7.31}$$

While a large number of choices are available for the prior density π_{21}, virtually all proponents of Bayesian analysis state that with no additional prior knowledge the most appropriate choice for the density is 1 (an improper prior). Therefore

$$\pi_{21}^*(\boldsymbol{\mu} \mid \boldsymbol{x},F,G) \propto exp[-(\hat{\boldsymbol{\theta}} - B\boldsymbol{\mu})'W(\hat{\boldsymbol{\theta}} - B\boldsymbol{\mu})/2]$$

$$\propto exp[-(\boldsymbol{\mu}'B'WB\boldsymbol{\mu} - 2\boldsymbol{\mu}'B'W\hat{\boldsymbol{\theta}})/2]$$

$$\boldsymbol{\mu} \mid \boldsymbol{x},F,G \sim N(\hat{\boldsymbol{\mu}},V_2)$$

$$\hat{\boldsymbol{\mu}} = V_2B'W\hat{\boldsymbol{\theta}} \qquad V_2 = (B'WB)^{-1} \tag{7.32}$$

where $\hat{\boldsymbol{\mu}}$ is the least squares estimate when the first two levels are combined into one.

We can now obtain the empirical Bayes posterior $\pi_{11}^*(\boldsymbol{\theta} \mid \boldsymbol{x},F,G)$ from (6.28). Once again this is a convolution of two normal distributions and so

$$\boldsymbol{\theta} \mid \boldsymbol{x},F,G \sim N(\boldsymbol{\theta}^*,V^*)$$

$$\boldsymbol{\theta}^* = E[(I - \Sigma W)\hat{\boldsymbol{\theta}} + \Sigma WB\boldsymbol{\mu} \mid \boldsymbol{x},F,G] = (I - \Sigma W)\hat{\boldsymbol{\theta}} + \Sigma WB\hat{\boldsymbol{\mu}} \tag{7.33}$$

$$V^* = Var[(I - \Sigma W)\hat{\boldsymbol{\theta}} + \Sigma WB\boldsymbol{\mu} \mid \boldsymbol{x},F,G] + E(V_1 \mid \boldsymbol{x},F,G)$$

$$= \Sigma WBV_2B'W\Sigma + V_1. \tag{7.34}$$

(7.33) is the customary empirical Bayes estimate for this general situation. Instead of a scalar credibility factor we now have a credibility matrix $Z = I - \Sigma W$. The empirical Bayes problem would now be to estimate the elements of F and G.

The last step is to use (6.26) to obtain $\pi_{22}^*(F,G \mid \boldsymbol{x})$. Before doing that we first must use (6.25), (7.21), (7.29), and (7.30) to obtain $m_{21}(\boldsymbol{x} \mid F,G)$:

$$m_{21}(\boldsymbol{x} \mid F,G) = \int m_1(\boldsymbol{x} \mid \boldsymbol{\mu},F,G)\pi_{21}(\boldsymbol{\mu} \mid F,G)d\boldsymbol{\mu}$$

$$= \int f_2(S \mid F)m_{11}(\hat{\boldsymbol{\theta}} \mid \boldsymbol{\mu},F,G)d\boldsymbol{\mu}$$

$$\propto f_2(S \mid F)|W|^{1/2} \int exp[-(\hat{\boldsymbol{\theta}} - B\boldsymbol{\mu})'W(\hat{\boldsymbol{\theta}} - B\boldsymbol{\mu})/2]d\boldsymbol{\mu}$$

$$\propto f_2(S \mid F)|W|^{1/2} \int exp[-\boldsymbol{\mu}'B'WB\boldsymbol{\mu}/2 + 2\boldsymbol{\mu}'B'W\hat{\boldsymbol{\theta}}/2 - \hat{\boldsymbol{\theta}}'W\hat{\boldsymbol{\theta}}/2]d\boldsymbol{\mu}$$

$$\propto f_2(S \mid F)|W|^{1/2}|V_2|^{1/2}exp[-\hat{\boldsymbol{\theta}}'W\hat{\boldsymbol{\theta}}/2 + \hat{\boldsymbol{\theta}}'WBV_2B'W\hat{\boldsymbol{\theta}}/2]$$

$$\propto f_2(S \mid F)|W|^{1/2}|V_2|^{1/2}exp[-\hat{\boldsymbol{\theta}}'W\hat{\boldsymbol{\theta}}/2 + 2\hat{\boldsymbol{\theta}}'WB\hat{\boldsymbol{\mu}}/2 + \hat{\boldsymbol{\mu}}'B'WB\hat{\boldsymbol{\mu}}/2]$$

$$\propto f_2(S \mid F)|W|^{1/2}|V_2|^{1/2}exp[-(\hat{\boldsymbol{\theta}} - B\hat{\boldsymbol{\mu}})'W(\hat{\boldsymbol{\theta}} - B\hat{\boldsymbol{\mu}})/2]$$

$$= f_2(S \mid F)m_{211}(\hat{\boldsymbol{\theta}} \mid F,G). \tag{7.35}$$

Then

$$\pi_{22}^*(F,G \mid \boldsymbol{x}) \propto m_{21}(\boldsymbol{x} \mid F,G)\pi_{22}(F,G) \propto f_2(S \mid F)m_{211}(\hat{\boldsymbol{\theta}} \mid F,G)\pi_{22}(F,G). \tag{7.36}$$

The three items of greatest interest are the posterior mean, variance, and covariance of the elements of $\boldsymbol{\theta}$. (6.31)—(6.34) provide the best way to get the first two of these:

$$m_i(F,G) = E(\theta_i \mid \boldsymbol{x},F,G) = \theta_i^*, \quad E(\theta_i \mid \boldsymbol{x}) = \int \int \theta_i^* \pi_{22}^*(F,G \mid \boldsymbol{x})dFdG \tag{7.37}$$

$$s_i(F,G) = Var(\theta_i \mid \boldsymbol{x},F,G) = v_{ii}^*,$$

$$Var(\theta_i \mid \boldsymbol{x}) = \int \int v_{ii}^* \pi_{22}^*(F,G \mid \boldsymbol{x})dFdG$$

$$+ \int \int (\theta_i^*)^2 \pi_{22}^*(F,G \mid \boldsymbol{x})dFdG - [E(\theta_i \mid \boldsymbol{x})]^2 \tag{7.38}$$

where θ_i^* is the ith element of $\boldsymbol{\theta}^*$ (7.33) and v_{ii}^* is the iith element of V^*

(7.34). The covariance can be obtained by using the same steps that led to (6.33):

$$Cov(\theta_i,\theta_j \mid \boldsymbol{x}) = \int\int v_{ij}^*\pi_{22}^*(F,G \mid \boldsymbol{x})dFdG$$

$$+ \int\int \theta_i^*\theta_j^*\pi_{22}^*(F,G \mid \boldsymbol{x})dFdG - E(\theta_i \mid \boldsymbol{x})E(\theta_j \mid \boldsymbol{x}). \qquad (7.39)$$

If the posterior density of one of the elements of $\boldsymbol{\theta}$ is desired, it is

$$\pi^*(\theta_i \mid \boldsymbol{x}) = \int\int \phi[(\theta_i - \theta_i^*)(v_{ii}^*)^{-1/2}]\pi_{22}^*(F,G \mid \boldsymbol{x})dFdG \qquad (7.40)$$

where ϕ is the standard normal *pdf.* In all of these integrals it should be noted that (7.36) provides the density $\pi_{22}^*(F,G \mid \boldsymbol{x})dFdG$ only up to a constant of proportionality. This constant will have to be found numerically by integrating the right hand side of (7.36).

The formulas for predicting the next value of X are easy to work out. Let the next value have conditional distribution $Y \mid \boldsymbol{\theta},F \sim N(\boldsymbol{a}'\boldsymbol{\theta},s)$ where s is a known function of the elements of F. From (6.36) we have

$$E(Y \mid \boldsymbol{x}) = \int\int \boldsymbol{a}'\boldsymbol{\theta}^*\pi_{22}^*(F,G \mid \boldsymbol{x})dFdG = \sum a_iE(\theta_i \mid \boldsymbol{x}). \qquad (7.41)$$

while from (6.37) we have

$$Var(Y \mid \boldsymbol{x}) = E(s \mid \boldsymbol{x}) + Var(\boldsymbol{a}'\boldsymbol{\theta} \mid \boldsymbol{x})$$

$$= \int\int s\pi_{22}^*(F,G \mid \boldsymbol{x})dFdG + \boldsymbol{a}'Var(\boldsymbol{\theta} \mid \boldsymbol{x})\boldsymbol{a}. \qquad (7.42)$$

At the end of Chapter 6, one more item of interest was mentioned. Begin by using (6.39) and $\pi_{21}(\boldsymbol{\mu} \mid F,G) \propto 1$ to write

$$\pi_{12}^*(\boldsymbol{\theta} \mid F,G) \propto |G|^{-1/2} \int exp[-(\boldsymbol{\theta} - B\boldsymbol{\mu})'G^{-1}(\boldsymbol{\theta} - B\boldsymbol{\mu})/2]d\boldsymbol{\mu}$$

$$= |G|^{-1/2}exp(-\boldsymbol{\theta}'G^{-1}\boldsymbol{\theta}/2) \int exp[-(\boldsymbol{\mu}'B'G^{-1}B\boldsymbol{\mu} - 2\boldsymbol{\mu}'B'G^{-1}\boldsymbol{\theta})/2]d\boldsymbol{\mu}$$

$$\propto |G|^{-1/2}|B'G^{-1}B|^{-1/2}$$

$$\times exp[-\boldsymbol{\theta}'G^{-1}\boldsymbol{\theta}/2 + \boldsymbol{\theta}'G^{-1}B(B'G^{-1}B)^{-1}B'G^{-1}\boldsymbol{\theta}/2]. \qquad (7.43)$$

Then use (6.39) to obtain

$$\pi^*(\boldsymbol{\theta},F,G \mid X) \propto |F|^{-1/2}|G|^{-1/2}|B'G^{-1}B|^{-1/2}$$
$$\times\ exp[-S/2-(\boldsymbol{\theta}-\hat{\boldsymbol{\theta}})'\Sigma^{-1}(\boldsymbol{\theta}-\hat{\boldsymbol{\theta}})/2-\boldsymbol{\theta}'G^{-1}\boldsymbol{\theta}/2$$
$$+\boldsymbol{\theta}'G^{-1}B(B'G^{-1}B)^{-1}B'G^{-1}\boldsymbol{\theta}/2]\pi_{22}(F,G) \tag{7.44}$$

The key formulas needed to obtain the quantities in (7.36)—(7.42) are repeated in the following two boxes:

Important vectors and matrices – version I

$$\Sigma=(A'F^{-1}A)^{-1} \qquad \hat{\boldsymbol{\theta}}=\Sigma A'F^{-1}\boldsymbol{x} \qquad S=(\boldsymbol{x}-A\hat{\boldsymbol{\theta}})'F^{-1}(\boldsymbol{x}-A\hat{\boldsymbol{\theta}})$$

$$W=(\Sigma+G)^{-1} \qquad \boldsymbol{\mu}_1=(I-\Sigma W)\hat{\boldsymbol{\theta}}+\Sigma WB\boldsymbol{\mu}$$

$$V_1=(\Sigma^{-1}+G^{-1})^{-1}=\Sigma-\Sigma W\Sigma$$

$$\hat{\boldsymbol{\mu}}=V_2B'W\hat{\boldsymbol{\theta}} \qquad V_2=(B'WB)^{-1}$$

$$\boldsymbol{\theta}^*=(I-\Sigma W)\hat{\boldsymbol{\theta}}+\Sigma WB\hat{\boldsymbol{\mu}} \qquad V^*=\Sigma WBV_2B'W\Sigma+V_1$$

Important distributions – version I

$$f_2(S \mid F) \propto |\Sigma|^{1/2}|F|^{-1/2}exp(-S/2)$$

$$\boldsymbol{\theta} \mid \boldsymbol{x},\boldsymbol{\mu},F,G \sim N(\boldsymbol{\mu}_1,V_1) \quad (\pi_1^*)$$

$$\boldsymbol{\mu} \mid \boldsymbol{x},F,G \sim N(\hat{\boldsymbol{\mu}},V_2) \quad (\pi_{21}^*) \qquad \boldsymbol{\theta} \mid \boldsymbol{x},F,G \sim N(\boldsymbol{\theta}^*,V^*) \quad (\pi_{11}^*)$$

$$m_{211}(\hat{\boldsymbol{\theta}} \mid F,G) \propto |W|^{1/2}|V_2|^{1/2}exp[-(\hat{\boldsymbol{\theta}}-B\hat{\boldsymbol{\mu}})'W(\hat{\boldsymbol{\theta}}-B\hat{\boldsymbol{\mu}})/2]$$

$$\pi_{22}^*(F,G \mid \boldsymbol{x}) \propto f_2(S \mid F)m_{211}(\hat{\boldsymbol{\theta}} \mid F,G)\pi_{22}(F,G)$$

There are many different ways in which these formulas can be written. Only one other will be presented here. It is patterned after Harville (1977) and has a particular advantage in that all the quantities are available even if the matrix A is of less than full rank. In this situation the matrix Σ is not defined and so all the subsequent version I formulas fall apart. There are two ways to resolve the issue, one is to use generalized inverses, while the other is to use an appropriate transformation of the vector $\boldsymbol{\theta}$. The first will not be covered in this monograph; the second will be illustrated in Chapter 8. Harville's approach succeeds by noting that the only key quantity that does not exist is $\hat{\boldsymbol{\theta}}$. But this quantity is not necessary to a Bayesian analysis (recall, it is the frequentist estimate and appeared in the formulas above out of convenience, not necessity). Intuitively, the Bayes solution exists because even though the model is overspecified, the prior is not. The formulas are summarized in the next two boxes. The key relationship in showing that they are equivalent to version I is $W = \Sigma^{-1}(\Sigma^{-1}+G^{-1})^{-1}G^{-1}$.[16]

Important vectors and matrices – version II

$$\Sigma^{-1} = A'F^{-1}A \qquad V_1 = (\Sigma^{-1}+G^{-1})^{-1}$$

$$V_2 = (B'\Sigma^{-1}V_1G^{-1}B)^{-1} \qquad \hat{\boldsymbol{\mu}} = V_2B'G^{-1}V_1A'F^{-1}\boldsymbol{x}$$

$$\boldsymbol{\theta}^* = B\hat{\boldsymbol{\mu}} + V_1A'F^{-1}(\boldsymbol{x} - AB\hat{\boldsymbol{\mu}}) \qquad V^* = V_1G^{-1}BV_2B'G^{-1}V_1 + V_1$$

Important distributions – version II

$$\boldsymbol{\mu} \mid \boldsymbol{x},F,G \sim N(\hat{\boldsymbol{\mu}},V_2) \quad (\pi_{21}^*) \qquad \boldsymbol{\theta} \mid \boldsymbol{x},F,G \sim N(\boldsymbol{\theta}^*,V^*) \quad (\pi_{11}^*)$$

$$\pi_{22}^*(F,G \mid \boldsymbol{x}) \propto \pi_{22}(F,G)|F|^{-1/2}|G|^{-1/2}|V_1|^{1/2}|V_2|^{1/2}$$
$$\times\ exp[-(\boldsymbol{x}'F^{-1}\boldsymbol{x} - \boldsymbol{x}'F^{-1}AV^*A'F^{-1}\boldsymbol{x})/2]$$

D. EXAMPLES – ANALYSIS

In this section the six models introduced in Section B will be analyzed using the formulas from Section C. Because the one-way model is

[16] The notation Σ^{-1} for the matrix $A'F^{-1}A$ is a bit improper since it is not the inverse of any matrix. Since the version II formulas do not require the inverse of this matrix, there is no problem in implementing them.

used the most, it will be presented in complete detail. Only the ultimate results will be presented for the other models.

1. One-way

$$A'F^{-1}A = diag(P_1,\ldots,P_k)/\sigma^2, \qquad P_i = \sum_{j=1}^{n_i} P_{ij}$$

$$\Sigma = diag(\sigma^2/P_1,\ldots,\sigma^2/P_k)$$

$$\hat{\theta}_i = \sum_{j=1}^{n_i} P_{ij}X_{ij}/P_i$$

$$S = \sum_{i=1}^{k}\sum_{j=1}^{n_i} P_{ij}(X_{ij}-\hat{\theta}_i)^2/\sigma^2 = WSS/\sigma^2$$

$$\Sigma + G = diag(\frac{\sigma^2}{P_1}+\tau^2,\ldots,\frac{\sigma^2}{P_k}+\tau^2)$$

$$W = diag(w_1,\ldots,w_k), \quad w_i = (\frac{\sigma^2}{P_i}+\tau^2)^{-1} = P_i/(\sigma^2+P_i\tau^2)$$

$$V_1 = diag\left[\frac{\sigma^2}{P_1}(1-\frac{\sigma^2}{P_1}w_1),\ldots,\frac{\sigma^2}{P_k}(1-\frac{\sigma^2}{P_k}w_k)\right]$$

$$V_2 = 1/\sum_{i=1}^{k} w_i = 1/w$$

$$\hat{\mu} = \sum_{i=1}^{k} w_i\hat{\theta}_i/w = \sum_{i=1}^{k}\frac{P_i\hat{\theta}_i}{\sigma^2+P_i\tau^2}\Big/\sum_{i=1}^{k}\frac{P_i}{\sigma^2+P_i\tau^2}$$

$$\theta_i^* = (1-\sigma^2 w_i/P_i)\hat{\theta}_i + (\sigma^2 w_i/P_i)\hat{\mu} = z_i\hat{\theta}_i + (1-z_i)\hat{\mu}$$

$$v_{ii}^* = \sigma^2 z_i/P_i + (1-z_i)^2/w$$

$$v_{ij}^* = (1-z_i)(1-z_j)/w \tag{7.45}$$

The credibility factor in the empirical Bayes formula for θ_i^* can be written

$$z_i = P_i/(P_i + \frac{\sigma^2}{\tau^2}).$$

This corresponds precisely to the Bühlmann estimate (5.6). The two key distributions are

$$f_2(S \mid F) \propto (\sigma^2)^{-(N-k)/2}\, exp(-WSS/2\sigma^2) \tag{7.46}$$

$$m_{211}(\hat{\boldsymbol{\theta}} \mid F,G) \propto (\prod_{i=1}^{k} w_i)^{1/2}(w)^{-1/2} exp[-\sum_{i=1}^{k} w_i(\hat{\theta}_i - \hat{\mu})^2/2]. \tag{7.47}$$

The posterior density of the variance components is

$$\pi^*(\sigma^2,\tau^2 \mid \boldsymbol{x})$$

$$\propto \frac{\pi_{22}(\sigma^2,\tau^2)(\sigma^2)^{-\frac{N-k}{2}}}{\left\{\sum_{i=1}^{k}\frac{P_i}{\sigma^2 + P_i\tau^2}\right\}^{1/2}\prod_{i=1}^{k}(\sigma^2 + P_i\tau^2)^{1/2}}\, exp\left[-\frac{WSS}{2\sigma^2} - \sum_{i=1}^{k}\frac{P_i(\hat{\theta}_i - \hat{\mu})^2}{2(\sigma^2 + P_i\tau^2)}\right].$$

$$\hat{\mu} = \sum_{i=1}^{k}\frac{P_i\hat{\theta}_i}{(\sigma^2 + P_i\tau^2)} \Big/ \sum_{i=1}^{k}\frac{P_i}{(\sigma^2 + P_i\tau^2)}. \tag{7.48}$$

Evaluation of the required integrals can be simplified if the following transformation is made:

$$\alpha = \sigma^2, \qquad \delta = \sigma^2/\tau^2. \tag{7.49}$$

The Jacobian for this transformation is α/δ^2. We also have $w_i = P_i\delta/\alpha(P_i + \delta)$ and so

$$\pi^*(\alpha,\delta \mid \boldsymbol{x})$$

$$\propto \frac{\pi_{22}(\alpha,\delta)\alpha^{-\frac{N-3}{2}}\delta^{\frac{k-5}{2}}}{\left\{\sum_{i=1}^{k}\frac{P_i}{P_i + \delta}\right\}^{1/2}\prod_{i=1}^{k}(P_i + \delta)^{1/2}}\, exp\left[-\frac{WSS}{2\alpha} - \sum_{i=1}^{k}\frac{P_i\delta(\hat{\theta}_i - \hat{\mu})^2}{2\alpha(P_i + \delta)}\right]. \tag{7.50}$$

Because the Jacobian has already been included, the prior density $\pi_{22}(\alpha,\delta)$ is obtained by applying the transformation to $\pi_{22}(\sigma^2,\tau^2)$ without introducing the Jacobian again. As a function of α (7.50) is an inverse gamma density (provided $\pi_{22}(\alpha,\delta)$ is a reasonable function of α) and so integrations with respect to α will be routine. This will leave a one-dimensional integration with respect to δ which can be done to a high degree of accuracy by adaptive Gaussian integration.

2. Two-way

Two versions of this model were presented in Section B. We begin with the simpler one ($\gamma^2 = 0$) in which the final estimates were constrained to be linear in the two factors. Begin by arranging the observations into a single vector $\boldsymbol{x}' = (X_{111},\ldots,X_{11T},X_{121},\ldots,X_{1JT},X_{211},\ldots,X_{IJT})$. Here there are I levels of the first classification variable, J levels of the second one, and observations are available for T years. The first level quantities are

$$\boldsymbol{\theta}' = (\alpha_1,\ldots,\alpha_I,\beta_1,\ldots,\beta_J) \text{ and } A = \begin{bmatrix} \mathbf{1} & \mathbf{0} & \cdots & \mathbf{0} & A_2 \\ \mathbf{0} & \mathbf{1} & \cdots & \mathbf{0} & A_2 \\ \vdots & \vdots & & \vdots & \vdots \\ \mathbf{0} & \mathbf{0} & \cdots & \mathbf{1} & A_2 \end{bmatrix}. \tag{7.51}$$

In the description of the matrix A the $\mathbf{1}$'s are vectors of length JT consisting of all ones. The $\mathbf{0}$'s are vectors of length JT consisting entirely of zeros. The $(JT \times J)$ matrices A_2 are each like the matrix in (7.4) with each $n_i = T$. The dimension of A is $IJT \times (I + J)$. This is an awkward way to describe this matrix. An example should make things clear. With $I = 2$, $J = 2$ and $T = 2$ the matrix is

$$A = \begin{bmatrix} 1 & 0 & 1 & 0 \\ 1 & 0 & 1 & 0 \\ 1 & 0 & 0 & 1 \\ 1 & 0 & 0 & 1 \\ 0 & 1 & 1 & 0 \\ 0 & 1 & 1 & 0 \\ 0 & 1 & 0 & 1 \\ 0 & 1 & 0 & 1 \end{bmatrix}. \tag{7.52}$$

At the second level the vector $\boldsymbol{\mu}$ has the single element μ while the $(I+J)\times 1$ matrix B is the vector with all elements equal to 1/2. The covariance G is diagonal with the first I diagonal elements τ_α^2 and last J diagonal elements τ_β^2. From here it would be a routine matter to apply the formulas from the previous section except that the matrix A is not of full rank and therefore many of the required inverses will not exist. This is not surprising as the model is overspecified in that an increase in each of the α's accompanied by a corresponding decrease in each of the β's leaves the first level means unchanged.

One approach is to use the version II formulas. Two of the matrices have a convenient expression:

$$\Sigma^{-1} = A'F^{-1}A = \begin{bmatrix} P_1 & P_{12} \\ P_{21} & P_2 \end{bmatrix} \frac{1}{\sigma^2} \tag{7.53}$$

$$A'F^{-1}\boldsymbol{x} = \begin{bmatrix} \sum\limits_{jt} P_{1jt}X_{1jt} \\ \vdots \\ \sum\limits_{jt} P_{Ijt}X_{Ijt} \\ \sum\limits_{it} P_{i1t}X_{i1t} \\ \vdots \\ \sum\limits_{it} P_{iJt}X_{iJt} \end{bmatrix} \frac{1}{\sigma^2}. \tag{7.54}$$

where P_1 is $I\times I$ diagonal with ith diagonal element

$P_{i\cdot\cdot} = \sum\limits_{jt} P_{ijt}$,

P_2 is $J\times J$ diagonal with jth diagonal element

$P_{\cdot j\cdot} = \sum\limits_{it} P_{ijt}$,

and P_{12} is $I\times J$ with ijth element

$P_{ij\cdot} = \sum\limits_{t} P_{ijt}$.

There is another approach that will work for any hierarchical linear model in which the first level design matrix is of less than full rank. Suppose A has dimension $N\times k$ and has rank $r<k$. Then A can always be

factored as the product of three matrices, $A = USE$ where U is $N \times r$, has rank r, and $U'U = I$, the identity matrix, S is $r \times r$ and is diagonal, while E is $r \times k$, has rank r, and $EE' = I$, the identity matrix. This is known as the singular value decomposition and is available in many computer packages. Let $C = US$ and so $A = CE$. Another way to obtain the factorization is to find the eigenvalues and eigenvectors of the matrix $A'A$. Let E be the matrix whose rows are the r eigenvectors that correspond to the non-zero eigenvalues. In either case the matrix C is not needed.

The next step is to insert an extra stage into the model as follows:

$$\begin{aligned}
&\boldsymbol{x} \mid \boldsymbol{\nu}, F \sim N(C\boldsymbol{\nu}, F) \\
&\boldsymbol{\nu} \mid \boldsymbol{\theta} \sim N(E\boldsymbol{\theta}, 0) \\
&\boldsymbol{\theta} \mid \boldsymbol{\mu}, G \sim N(B\boldsymbol{\mu}, G) \\
&\pi_2(\boldsymbol{\mu}, F, G)
\end{aligned} \tag{7.55}$$

Collapsing the first two levels restores the original three-level model. The trick is to collapse the two middle levels to yield:

$$\begin{aligned}
&\boldsymbol{x} \mid \boldsymbol{\nu}, F \sim N(C\boldsymbol{\nu}, F) \\
&\boldsymbol{\nu} \mid \boldsymbol{\mu}, G \sim N(EB\boldsymbol{\mu}, EGE') \\
&\pi_2(\boldsymbol{\mu}, F, G)
\end{aligned} \tag{7.56}$$

This is a full-rank model and can be analyzed with the version I formulas. Of course the vector to be estimated is now $\boldsymbol{\nu}$ which will carry no useful interpretation. However, the vector $C\boldsymbol{\nu} = AE'\boldsymbol{\nu}$ will be the correct vector for predicting future claims. It is not possible to uniquely recover the matrix of interest $\boldsymbol{\theta}$ since any solution to the equation $\boldsymbol{\nu} = E\boldsymbol{\theta}$ will suffice. By adding some additional constraints a unique solution can be found.

For the two-way model with $I = J = T = 2$ the matrix A as given in (7.52) can be factored as

$$E = \begin{bmatrix} -.5 & -.5 & -.5 & -.5 \\ -.5493 & .5493 & .4453 & -.4453 \\ .4453 & -.4453 & .5493 & -.5493 \end{bmatrix}, \quad \text{and}$$

$$C = \begin{bmatrix} -1 & -.104 & .995 \\ -1 & -.104 & .995 \\ -1 & -.995 & -.104 \\ -1 & -.995 & -.104 \\ -1 & .995 & .104 \\ -1 & .995 & .104 \\ -1 & .104 & -.995 \\ -1 & .104 & -.995 \end{bmatrix}$$

Suppose that the true values of the parameters were, $\alpha_1 = 2$, $\alpha_2 = 3$, $\beta_1 = 4$, and $\beta_2 = 2$. The second level vector is $\boldsymbol{\theta} = (2,\ 3,\ 4,\ 2)'$. The corresponding vector $\boldsymbol{\nu}$ is $(-5.5,\ 1.4399,\ .6532)'$. To recover $\boldsymbol{\theta}$ from $\boldsymbol{\nu}$ it is necessary to introduce a constraint. Add one zero to the end of $\boldsymbol{\nu}$ and add the appropriate row to the matrix E to obtain the following equation for $\boldsymbol{\theta}$:

$$\begin{bmatrix} -.5 & -.5 & -.5 & -.5 \\ -.5493 & .5493 & .4453 & -.4453 \\ .4453 & -.4453 & .5493 & -.5493 \\ 1 & 1 & 0 & 0 \end{bmatrix} \boldsymbol{\theta} = \begin{bmatrix} -5.5 \\ .4399 \\ .6532 \\ 0 \end{bmatrix}.$$

The restriction requested here is that the α's add to zero. The unique solution to this equation is the vector $\boldsymbol{\theta}^+ = (-.5,\ .5,\ 6.5,\ 4.5)'$. While this does not equal the original vector it is easy to see that $A\boldsymbol{\theta}^+ = A\boldsymbol{\theta}$. Call the augmented E matrix E^* and then inferences about $\boldsymbol{\theta}$ can be obtained by noting that it is just $(E^*)^{-1}\boldsymbol{\nu}^+$ where $\boldsymbol{\nu}^+$ is the vector $\boldsymbol{\nu}$ with the zero added on. A more reasonable constraint might be that the exposure weighted sums of the α's and β's are equal. If no particular constraints are required the vector $E'\boldsymbol{\nu}$ will suffice.

A look at (7.56) indicates that the only other items that need to be evaluated are $B^* = EB$ and $G^* = EGE'$. The model is now

$$\begin{aligned} &\boldsymbol{x} \mid \boldsymbol{\nu}, F \sim N(C\boldsymbol{\nu}, F) \\ &\boldsymbol{\nu} \mid \mu, G^* \sim N(B^*\mu, G^*) \\ &\pi_2(\mu, F, G^*). \end{aligned} \tag{7.57}$$

A small example will be introduced here to illustrate the two methods of attacking the two-way model. Continuing to use $I = J = T = 2$ let the observations be

i	1	1	1	1	2	2	2	2
j	1	1	2	2	1	1	2	2
t	1	2	1	2	1	2	1	2
P_{ijt}	1	2	3	4	3	5	4	5
X_{ijt}	7	6	5	4	2	3	1	0

Further assume $\sigma^2 = 1$, $\tau_\alpha^2 = 2$, and $\tau_\beta^2 = 3$. A few of the intermediate calculations are displayed below:

$$\hat{\boldsymbol{\nu}} = \begin{bmatrix} -3.2967 \\ -.91765 \\ 2.8182 \end{bmatrix}, \quad B^* = \begin{bmatrix} -1 \\ 0 \\ 0 \end{bmatrix}, \quad G^* = \begin{bmatrix} 2.5 & 0 & 0 \\ 0 & 2.3966 & .48920 \\ 0 & .48920 & 2.6034 \end{bmatrix}$$

$$\hat{\mu} = 3.27007, \qquad \boldsymbol{\nu}^* = \begin{bmatrix} -3.2701 \\ -.87245 \\ 2.7083 \end{bmatrix}, \quad \boldsymbol{\theta}^* = \begin{bmatrix} 3.4370 \\ .06647 \\ 2.6174 \\ .41932 \end{bmatrix}.$$

To see that the version I formulas do not require the matrix C, insert $EE' = I$ where appropriate to obtain the following:

$$\begin{aligned} \Sigma^{-1} &= C'F^{-1}C = EA'F^{-1}AE' \\ \hat{\boldsymbol{\nu}} &= \Sigma EA'F^{-1}\boldsymbol{x} \\ S &= \boldsymbol{x}'F^{-1}\boldsymbol{x} - \hat{\boldsymbol{\nu}}'\Sigma^{-1}\hat{\boldsymbol{\nu}}. \end{aligned} \tag{7.58}$$

(7.53) and (7.54) will help in evaluating these quantities. The first term in the expression for S is just $\sum_{ijt} P_{ijt}X_{ijt}^2$.

For doing the factorization note that in general the matrix $A'A$ is

$$A'A = \begin{bmatrix} JT & & & T & \cdots & T \\ & \ddots & & \vdots & \ddots & \vdots \\ & & JT & T & \cdots & T \\ T & \cdots & T & IT & & \\ \vdots & \ddots & \vdots & & \ddots & \\ T & \cdots & T & & & IT \end{bmatrix}. \tag{7.59}$$

The calculations were also done using the version II formulas. The value of $\hat{\mu}$ is the same as above, but a different value of $\boldsymbol{\theta}^*$ obtains. It is $(3.3203, -.05023, 2.7341, .53602)'$. While at first this appears to be a different answer it should be noted that adding .1167 to each of the first two values and subtracting it from each of the last two will produce the version I answer. The estimates for the class means will be the same in either case. When version II is being used the restrictions that produce a unique answer are implicit. The analyst is free to make any adjustments at the end if a specific restriction is contemplated.

The second version of the two-way model is a bit more complex. The matrix A is the same as in the one-way model with each vector of length T. The matrix B is $IJ \times 1$ and is all ones. The vector $\boldsymbol{\mu}$ is the constant μ. The complications are all in the $IJ \times IJ$ matrix G. It can be written (using (7.9))

$$G = \begin{bmatrix} G_1 & 0 & \cdots & 0 \\ 0 & G_1 & \ddots & \vdots \\ \vdots & \ddots & \ddots & 0 \\ 0 & \cdots & 0 & G_1 \end{bmatrix} + \begin{bmatrix} G_2 & \cdots & \cdots & G_2 \\ \vdots & \ddots & & \vdots \\ \vdots & & \ddots & \vdots \\ G_2 & \cdots & \cdots & G_2 \end{bmatrix} \tag{7.60}$$

where the $J \times J$ matrix G_2 is diagonal with each diagonal element equal to τ_β^2 and the $J \times J$ matrix G_1 has diagonal elements $\gamma^2 + \tau_\alpha^2$ and off-diagonal elements τ_α^2. Very little simplification is available. The least squares solution is

$$\hat{\boldsymbol{\theta}} = (\theta_{11}, \theta_{12}, \ldots, \theta_{IJ})' \text{ with } \hat{\theta}_{ij} = \sum_t P_{ijt} X_{ijt} / P_{ij\cdot} \text{ and } P_{ij\cdot} = \sum_t P_{ijt}.$$

Also,

$$WSS = \sum_{ijt} P_{ijt}(X_{ijt} - \hat{\theta}_{ij})^2$$

and the matrix Σ is diagonal with typical element $\sigma^2/P_{ij\cdot}$.

When applying the version I formulas the only problem is obtaining the inverse of $\Sigma + G$. First, note that this matrix has the form

$$W^{-1} = \begin{bmatrix} G_{11} & 0 & \cdots & 0 \\ 0 & G_{12} & & \vdots \\ \vdots & & \ddots & 0 \\ 0 & \cdots & 0 & G_{1I} \end{bmatrix} + \begin{bmatrix} G_2 & \cdots & \cdots & G_2 \\ \vdots & \ddots & & \vdots \\ \vdots & & \ddots & \vdots \\ G_2 & \cdots & \cdots & G_2 \end{bmatrix}. \tag{7.61}$$

In general the inverse of a matrix of the form (7.61) can be written in blocks with the ijth block equal to

$$\delta_{ij} G_{1i}^{-1} - G_{1i}^{-1}[I_J + G_2 \sum_{k=1}^{I} G_{1k}^{-1}]^{-1} G_2 G_{1j}^{-1} \tag{7.62}$$

where $\delta_{ij} = 1$ if $i = j$ and $\delta_{ij} = 0$ otherwise and I_J is the $J \times J$ identity matrix. All the inversions are of matrices of dimension $J \times J$ and so it is a good strategy to select the two factors so that the second one has the smallest number of levels.

The small data set introduced in this section was also analyzed with this model. The parameter γ^2 was set equal to 5. The least squares solution was $\hat{\boldsymbol{\theta}} = (6.3333, 3.8571, 2.6250, 0.4444)'$. Note that this vector has a different interpretation than the same vector from the simpler model. In that case the elements were estimates of the additive factors while here they are estimates of the cell means. The vectors have the same dimension because $I + J = IJ$ in this case. The posterior mean is $\boldsymbol{\theta}^* = (6.2307, 3.8459, 2.6377, 0.4762)'$. The first thing to note is that the four means do not follow an additive model. When the simpler additive model was fit the vector was (found by multiplying either of the two solutions by the matrix A) $(6.0544, 3.8563, 2.6839, 0.4858)'$. In three of the four cases the new model produced a result that was between the old model and the least squares estimate. A reason this was not true in all four cases is that all the models are such that the exposure weighted sum of the answer must equal the exposure weighted sum of the original data. (It is 71 for all three vectors in this paragraph.)

3. Linear Trend

This model can be placed in the *HNLM* framework by first arranging the observations as was done in the one-way model. That is, $\boldsymbol{x}' = (X_{11},\ldots,X_{1t},X_{21},\ldots,X_{kt})$. The remaining elements are then as follows:

$$\boldsymbol{\theta}' = (\alpha_1,\beta_1,\ldots,\alpha_k,\beta_k), \quad A = diag(a_1,\ldots,a_k), \quad a_k = \begin{bmatrix} 1 & 1 \\ \vdots & \vdots \\ 1 & t \end{bmatrix},$$

$$F = diag(\sigma^2/P_{11},\ldots,\sigma^2/P_{kt})$$

$$\boldsymbol{\mu}' = (\mu_1,\mu_2), \quad B = \begin{bmatrix} 1 & 0 \\ 0 & 1 \\ \vdots & \vdots \\ 1 & 0 \\ 0 & 1 \end{bmatrix},$$

$$G = diag(g,\ldots,g), \quad g = \begin{bmatrix} \tau_\alpha^2 & \tau_{\alpha\beta} \\ \tau_{\alpha\beta} & \tau_\beta^2 \end{bmatrix}. \tag{7.63}$$

The first few calculations are easy to do. Begin by defining a few sums and weighted averages:

$$P_i = \sum_{j=1}^{t} P_{ij}, \quad P_{i+} = \sum_{j=1}^{t} jP_{ij}, \quad P_{i++} = \sum_{j=1}^{t} j^2 P_{ij}$$

$$\overline{X}_i = \sum_{j=1}^{t} P_{ij}X_{ij}/P_i, \quad \overline{X}_{i+} = \sum_{j=1}^{t} jP_{ij}X_{ij}/P_{i+}. \tag{7.64}$$

The least squares estimates and the two basic matrices are then:

$$\hat{\boldsymbol{\theta}} = (\hat{\alpha}_i, \hat{\beta}_1, \ldots, \hat{\alpha}_k, \hat{\beta}_k)'$$

$$\hat{\beta}_i = \frac{P_i P_{i+}(\overline{X}_{i+} - \overline{X}_i)}{P_i P_{i++} - P_{i+}^2}, \quad \hat{\alpha}_i = \overline{X}_i - \hat{\beta}_i P_{i+}/P_i \tag{7.65}$$

$$\Sigma = diag(s_1, \ldots, s_k), \quad s_i = \frac{\sigma^2}{p_i}\begin{bmatrix} P_{i++} & -P_{i+} \\ -P_{i+} & P_i \end{bmatrix}, \quad p_i = P_i P_{i++} - P_{i+}^2 \tag{7.66}$$

$$W = diag(w_1, \ldots, w_k), \quad w_i = \frac{1}{d_i}\begin{bmatrix} \sigma^2 P_i/p_i + \tau_\beta^2 & \sigma^2 P_{i+}/p_i - \tau_{\alpha\beta} \\ \sigma^2 P_{i+}/p_i - \tau_{\alpha\beta} & \sigma^2 P_{i++}/p_i + \tau_\alpha^2 \end{bmatrix}$$

$$d_i = \sigma^2(\sigma^2 + \tau_\alpha^2 P_i + \tau_\beta^2 P_{i++} + 2\tau_{\alpha\beta} P_{i+})/p_i + (\tau_\alpha^2 \tau_\beta^2 - \tau_{\alpha\beta}^2) \tag{7.67}$$

$$\hat{\boldsymbol{\mu}} = \left(\sum w^i\right)^{-1} \sum w_i (\hat{\alpha}_i, \hat{\beta}_i)' \tag{7.68}$$

$$(\boldsymbol{\mu}_1)_i = (I - s_i w_i)(\hat{\alpha}_i, \hat{\beta}_i)' + s_i w_i \hat{\boldsymbol{\mu}} \tag{7.69}$$

$$\begin{aligned} \pi_{22}^*(\sigma^2, \tau_\alpha^2, \tau_\beta^2, \tau_{\alpha\beta} \mid \boldsymbol{x}) &\propto (\sigma^2)^{-kt/2}\left(\prod p_i\right)^{1/2}\left(\prod d_i\right)^{-1/2}\left|\sum w_i\right|^{-1/2} \\ &\times exp\left\{-S/2 - \sum [(\hat{\alpha}_i, \hat{\beta}_i)' - \hat{\boldsymbol{\mu}}]' w_i [(\hat{\alpha}_i, \hat{\beta}_i)' - \hat{\boldsymbol{\mu}}]\right\} \\ &\times \pi_{22}(\sigma^2, \tau_\alpha^2, \tau_\beta^2, \tau_{\alpha\beta}) \end{aligned} \tag{7.70}$$

$$S = \sum [\boldsymbol{x}_i - a_i(\hat{\alpha}_i, \hat{\beta}_i)'] F_i^{-1} [\boldsymbol{x}_i - a_i(\hat{\alpha}_i, \hat{\beta}_i)'] \tag{7.71}$$

where $\boldsymbol{x}_i$ and F_i^{-1} refer to the ith block of t items of the respective vector and matrix.

There is not much useful simplification that can be done after this point. It should be noted that when integrating over the various parameters the support for G is the set of values $\tau_\alpha^2 \tau_\beta^2 - \tau_{\alpha\beta}^2 > 0$. This insures that G is positive definite. This is another *example* of a model for which the exact calculation of posterior quantities will be difficult. Recall from Chapter 4 that it is also possible to express a linear trend model in the

Kalman filter form.

4. Kalman Filter

We continue here with the autoregressive model introduced in Section B. Running the Kalman filter separately on each class yields the distribution at time n

$$\tilde{\boldsymbol{\mu}}_{in} \sim N(\boldsymbol{\mu}_{in}, C_{in}). \tag{7.72}$$

View this as the first stage of a *HNLM* with the second and third stages given by

$$\boldsymbol{\mu}_{in} \sim N(B^*\boldsymbol{\mu}_n^*, G^*)$$

$$\pi(\boldsymbol{\mu}_n^*, G^*). \tag{7.73}$$

The only difference here is that the first level variance (F in the usual formulation) is known and as well the first level design matrix (A) is the identity matrix. Difficulties in implementing the analysis will depend on the complexity of the structure of G^*. For both models presented earlier in this chapter it seems best to be as general as possible:

$$B^* = \begin{bmatrix} 1 \\ 1 \end{bmatrix} \quad \boldsymbol{\mu}^* = \mu^* \quad G^* = \begin{bmatrix} g_1^2 & g_{12} \\ g_{12} & g_2^2 \end{bmatrix} \tag{7.74}$$

where $g_1^2 g_2^2 - g_{12}^2 > 0$ is required to ensure a positive definite matrix.

5. Graduation

Some adjustments to the standard analysis are required due to $\boldsymbol{\mu}$ being known. Using the version I formulas the relevant posterior is

$$\boldsymbol{\theta} \mid \boldsymbol{x}, \tau^2 \sim N(\boldsymbol{\mu}_1, V_1)$$

$$\boldsymbol{\mu}_1 = (F^{-1} + K'K/\tau^2)^{-1} F^{-1}\boldsymbol{x} \qquad V_1 = (F^{-1} + K'K/\tau^2)^{-1}. \tag{7.75}$$

This is the same posterior distribution as given in (4.5) and $\boldsymbol{\mu}_1$ is the usual Whittaker graduation. To estimate τ^2 another posterior distribution is required:

$$\pi_{22}^*(\tau^2 \mid \boldsymbol{x}) \propto |W|^{1/2} exp(-\boldsymbol{x}'W\boldsymbol{x}/2) \qquad W = [F + \tau^2(K'K)^{-1}]^{-1}. \tag{7.76}$$

E. PRIOR DISTRIBUTIONS

The final element to be specified is the prior distribution of the variances, $\pi_{22}(F,G)$. In a hierarchical Bayesian analysis a noninformative prior is a reasonable choice. This is because all relevant prior information should have been incorporated in the specification of the distributions at the first two levels. This was seen in the various models displayed in Section B.

Two choices for a noninformative prior will be offered here. The first has the advantage of being elementary. It is $\pi_{22}(F,G) \propto 1$. If the two covariance matrices are parametrized in terms of variances and correlations then the uniform distribution for the correlations is certainly reasonable as it will be a proper prior distribution. Recall from Chapter 2 that this prior (along with others) has the desirable property of being relatively scale invariant. On the other hand, this prior has the undesirable property of overemphasizing large values of the variance.

Additional problems arise when there are multiple parameters. The Jeffreys prior in this case is the square root of the determinant of the information matrix. Besides the difficulty in evaluating this quantity, there is evidence that in even the most elementary cases it is not satisfactory (Berger, 1985, p. 88). In the *HNLM* the density to look at is

$$m_1(\boldsymbol{x} \mid \boldsymbol{\mu}, F, G) \sim N(AB\boldsymbol{\mu}, AGA' + F)$$

(7.28). For a normal distribution with arbitrary covariance C the Jeffreys prior is

$$|C|^{-(m+1)/2}$$

where m is the dimension of the random variable (Box and Tiao, 1973, p. 426). As was seen in Section B, our models do not use arbitrary covariance matrices. For example, in the one-way model $AGA' + F$ is a function of two random variables, σ^2 and τ^2. Even so, it would not be unreasonable to use the determinant as a starting point. From (7.29) we see that the determinant is also $|F|/|\Sigma||W|$. For the one-way model this is proportional to

$$(\sigma^2)^{N-k}\Pi(\sigma^2 + P_i\tau^2).$$

This should then be taken to an appropriate negative power. A reasonable choice for a noninformative prior is

$$(\sigma^2)^{-1}/[\Pi(\sigma^2 + P_i\tau^2)]^{1/k}.$$

When the P_i are identical this matches the prior recommended by Box and Tiao (1973, p. 251).

An alternative in the multiparameter case is to use the reciprocal of the geometric mean of the diagonal elements of $AGA' + F$ raised to a power equal to the number of parameters. This ignores any correlation structure (or, equivalently, places a prior of 1 on the correlations). This is likely to be much easier to obtain than the three determinants used above.

Examples illustrating the effects of using these different priors are presented in the next Chapter.

F. MODEL SELECTION AND EVALUATION

When several models are available for the same situation it becomes necessary to have a mechanism for determining the one that is most appropriate. This is especially true for nested models. For example, the one-way model is a special case of each of the other models presented in Section B. Even for the one-way model there are a number of assumptions that should be checked: normality and constant variance being the most crucial. In this Section the graphical techniques introduced in Chapter 2 will be re-introduced for use with the *HNLM*.

The standard technique for evaluating the assumptions used in a regression analysis is to plot the residuals against the fitted values. For the *HNLM* this must be checked at both the first and second levels as normal models have been assumed for both. Box (1980 and 1983) also suggests looking at the marginal distribution as found from the estimated prior distribution and then seeing if the observations appear to be a random sample from that distribution.

One way to do this is to look at the distributions from the individual levels conditioned on the estimates of the various parameters:

$$\textit{First level} — \boldsymbol{x} \sim N(A\hat{\boldsymbol{\theta}},\hat{F}) \tag{7.77}$$

$$\textit{Second level} — \hat{\boldsymbol{\theta}} \sim N(B\hat{\boldsymbol{\mu}},\hat{\Sigma}+\hat{G}) \tag{7.78}$$

$$\textit{Overall} — \boldsymbol{x} \sim N(AB\hat{\boldsymbol{\mu}},\hat{F}+A\hat{G}A'). \tag{7.79}$$

Any of the estimates of F and G can be used here but for the normal distributions to make sense it would be best to use the least squares estimates of $\boldsymbol{\mu}$ and $\boldsymbol{\theta}$. To check out the assumptions the values must be standardized. That is, subtract the estimate of the mean and then pre-multiply by an appropriate matrix. This matrix can be found by first factoring the inverse of the covariance matrix as $C'C$ (the Choleski factorization is one way to do this). Pre-multiplying by C will complete the standardization.

Once the standardized residuals have been obtained they can be used in two ways. The first is to check for normality. While a formal test could be conducted it is sufficient to look at a plot. One plot would be the usual histogram which could then be compared to the standard normal density function. Another is to plot the ordered values on normal probability paper. That is, if $x_{(1)} < \ldots < x_{(n)}$ are the ordered standardized values, the ith point to plot is

$$(i,\Phi^{-1}(x_{(i)}))$$

where Φ is the standard normal distribution function. If the values are from a standard normal distribution these points will lie close to a straight line.

The second item to check is the variance assumption. While there are many ways in which the model could be incorrect, the most likely is that the variance bears some relationship to the size of the losses. The appropriate way to investigate this is to plot the standardized residuals against the fitted observations (this can only be done for level one where $\hat{\boldsymbol{\theta}}$ is the fitted observations) or against the observations themselves ($\hat{\boldsymbol{\theta}}$ for level two and $\boldsymbol{x}$ for the overall model).

Both of these items will be checked for an example with the one-way model in the *next* Chapter.

The Schwartz Bayesian Criterion is available for comparing two models. The key is the distribution in (7.79). Insert the estimates of the variance terms in this density and take the logarithm. It is

$$-[ln \mid \hat{F} + A\hat{G}A' \mid + (\boldsymbol{x} - AB\hat{\boldsymbol{\mu}})(\hat{F} + A\hat{G}A')^{-1}(\boldsymbol{x} - AB\hat{\boldsymbol{\mu}})]/2. \quad (7.80)$$

Complete the calculation of the SBC by subtracting $pln(n/2\pi)$ where p is the number of estimated parameters and n is the sample size.

8. EXAMPLES

In this Chapter a number of data sets will be introduced. Then the credibility models from the previous Chapter will be analyzed.

A. DATA SETS

The first data set is artificial and is designed to provide some benchmarks. In particular, it is selected so that the exact solutions can be obtained, something that is not possible in most practical situations. It is a one-way model with $k = 10$ groups and $n_i = n = 5$ observations from each group. The exposures are $P_{ij} = 1$ for all $N = 50$ observations. This last feature makes computation especially simple. Normal distributions will be assumed for the two levels with the first level variance $\sigma^2 = 100$ and the second level variance $\tau^2 = 225$. The overall mean is $\mu = 100$. Another advantage here is that since the data have been simulated the true values of $\theta_1, \ldots, \theta_{10}$ will be known. The 50 observations are presented in Table 8.1.

Two other values of interest are $WSS = 3382.82$ and $\hat{\mu} = 103.1842$ (The formula for $\hat{\mu}$ in (7.48) simplifies since $P_i = 5$ and $w_i = 5/(\sigma^2 + 5\tau^2)$ for all i). This allows (7.50) to be written in a reasonably concise manner:

$$\pi^*(\alpha,\delta \mid X) \propto \pi_{22}(\alpha,\delta)\alpha^{-23.5}\delta^{2.5}(5+\delta)^{-4.5}$$

$$\times exp[-3382.82/2\alpha - 5(737.703)\delta/2\alpha(5+\delta)]. \tag{8.1}$$

The other quantities of interest are

$$z_i = \frac{5}{(5+\delta)}$$

$$\theta_i^* = z_i\overline{X}_i + (1-z_i)103.1842 = \frac{5}{5+\delta}(\overline{X}_i - 103.1842) \ + \ 103.1842$$

$$v_{ii}^* \ = \ \frac{\alpha(50+\delta)}{50(5+\delta)}, \quad v_{ij}^* \ = \ \frac{\alpha\delta}{50(5+\delta)}. \tag{8.2}$$

Both (8.1) and (8.2) have employed the transformation $\alpha = \sigma^2$, $\delta = \sigma^2/\tau^2$. Note that all the integrations with respect to α will be easy as they are just inverse gamma densities. The integrals with respect to δ will have to be done numerically.

Table 8.1

Data Set 1 — Simulated values, one way model

i	X_{i1}	X_{i2}	X_{i3}	X_{i4}	X_{i5}	θ_i	$\overline{X}_i$
1	124.93	110.67	106.93	104.05	101.60	111.97	109.636
2	97.16	89.28	102.88	111.10	105.78	98.87	101.240
3	103.57	86.82	92.49	87.99	83.80	90.98	90.934
4	119.53	125.92	98.05	117.57	94.17	108.44	111.048
5	95.68	110.43	83.59	110.54	101.51	99.06	100.350
6	102.04	93.60	106.12	98.70	108.55	95.44	101.802
7	112.71	101.64	106.50	111.71	100.59	105.09	106.630
8	119.16	111.54	127.24	115.02	115.06	111.99	117.604
9	81.71	90.45	91.51	84.74	88.95	95.99	87.472
10	102.51	123.81	113.83	94.97	90.51	105.60	105.126

As mentioned, Data Set 1 will be used to test the accuracy of some of the approximations. It can also serve as a practice set for the interested reader who might be trying out some of the methods suggested here. The next example is also a one-way model but is closer to what might be found in practice. The observations are frequency counts on workers' compensation insurance. The data were collected from 133 occupation classes over a seven year period. Unlike Data Set 1 the observations do not have equal exposure. The exposures here are scaled payroll totals that have been inflated to represent constant dollars. These observations were taken in one state.[17] The results from the first two classes are given in Table 8.2. The complete data set appears in the Appendix. In this Chapter the various numerical techniques will be applied using the Normal model introduced in Chapter 7. In Chapter 9 a more realistic model will be presented and these data will be analyzed again using that model. The observations Y_{ij} are frequency counts for incidents of permanent partial disability.

Table 8.2

Data Set 2 — Workers' Compensation Frequencies

Class(i)	*Year(j)*	P_{ij}	Y_{ij}
1	1	32.322	1
1	2	33.779	4
1	3	43.548	3
1	4	46.686	5
1	5	34.713	1
1	6	32.857	3
1	7	36.600	4
2	1	45.995	3
2	2	37.888	1
2	3	34.581	0
2	4	28.298	0
2	5	45.265	2
2	6	39.945	0
2	7	39.322	4
⋮	⋮	⋮	⋮

The third data set contains workers' compensation loss ratios for the same seven year period. This time the exposures have not been adjusted for inflation. This will allow us to check out the model with linear trend. This data set will also be used for the parameter uncertainty (Kalman filter) model. Only losses relating to permanent partial disability were recorded. Both payroll and loss values were divided by 10^7 to make the scale more reasonable. The state is not the same one used for Data Set 2. The first few values appear in Table 8.3 while the complete data set appears in the Appendix.

The fourth data set is used to provide an illustration of the two-way model. To keep matters compact, this illustration uses only 25 of the

[17] These data were provided by the National Council on Compensation Insurance. At their request neither the state nor the years can be identified and the numbers identifying the classes do not reflect any NCCI coding.

133 occupation classes but now covers 10 different states. Again seven years of frequency data are available on permanent partial disability. As in Data Set 2, the exposures have already been adjusted for inflation.

Table 8.3

Data Set 3 — Worker's Compensation Pure Premiums

Class(i)	*Year(j)*	*Payroll*	*Loss*	*Pure Premium*
1	1	2.1798086	.0538707	0.02471
1	2	2.2640528	.0439184	0.01940
1	3	2.2572010	.1059775	0.04695
1	4	2.4789710	.0560013	0.02259
1	5	2.5876764	.1004997	0.03884
1	6	2.8033613	.1097314	0.03914
1	7	2.2525887	.0609833	0.02707
2	1	1.2004031	.0270222	0.02251
2	2	1.2713178	.0229566	0.01806
2	3	1.3596610	.0596850	0.04390
2	4	1.4811727	.0196539	0.01327
2	5	1.2774073	.0134248	0.01051
2	6	2.0245789	.0489312	0.02417
2	7	2.4242468	.0418218	0.01725
⋮	⋮	⋮	⋮	⋮

B. ANALYSES

1. One-way Model, Data Set 1

The first example uses Data Set 1. An examination of (8.2) indicates that integrals with respect to α can be done analytically but those with respect to δ cannot. Do this first integral by writing

$$I_t(\delta) = \int_0^\infty \alpha^{t-23.5}\delta^{2.5}(5+\delta)^{-4.5} exp\left[-\frac{3382.82}{2\alpha} - \frac{5(737.703)\delta}{2\alpha(5+\delta)}\right] d\alpha$$

$$= \delta^{2.5}(5+\delta)^{-4.5}\left[1691.41 + \frac{1844.2575\delta}{5+\delta}\right]^{-(22.5-t)} \Gamma(22.5-\text{t}). \qquad (8.3)$$

For this illustration the simple prior $\pi_{22}(\alpha,\delta) = 1$ will be used. In that case the constant of integration is

$$c = \int_0^\infty I_0(\delta)d\delta.$$

Adaptive Gaussian integration produced the value $exp(-178.4856)\Gamma(22.5)$. For obtaining point estimates the quantity of interest is

$$E(z_i \mid \boldsymbol{X}) = \int_0^\infty 5(5+\delta)^{-1} I_0(\delta)d\delta/c$$

$$= exp(-178.6818)\Gamma(22.5)/c = exp(-.1962) = .8218.$$

The posterior means are

$$.8218\overline{X}_i + .1782(103.1842) = .8218\overline{X}_i + 18.387. \tag{8.4}$$

The estimates are given in Table 8.4.

Table 8.4

Data Set 1 — Posterior means, variances, and confidence intervals

i	*Mean*	*Variance*	*Confidence interval*	
1	108.486	16.075	101.891	115.081
2	101.586	15.646	95.079	108.093
3	93.117	17.304	86.274	99.960
4	109.646	16.304	103.004	116.288
5	100.855	15.694	94.338	107.371
6	102.048	15.624	95.546	108.550
7	106.016	15.737	99.490	112.540
8	115.034	17.961	108.062	122.005
9	90.271	18.402	83.215	97.328
10	104.780	15.645	98.273	111.286

We might also be interested in the posterior variances. From (7.38) they are

$$E(v_{ii}^* \mid \boldsymbol{x}) + Var(\theta_i^* \mid \boldsymbol{x})$$

$$= E[.02\alpha(50+\delta)(5+\delta)^{-1} \mid \boldsymbol{x}]$$

$$+ (\overline{X}_i - 103.1842)^2 Var[5(5+\delta)^{-1} \mid \boldsymbol{x}]. \tag{8.5}$$

The expectation is

$$.02 \int_0^\infty (50+\delta)(5+\delta)^{-1} I_1(\delta) d\delta / c = exp(-172.6701)\Gamma(21.5)/c$$

$$= exp(5.8155)/21.5 \ = 15.6027.$$

The variance can be found by first obtaining the second moment as

$$25 \int_0^\infty (5+\delta)^{-2} I_0(\delta) d\delta / c$$

$$= exp(-178.8614)\Gamma(22.5)/c = exp(-.3758) = .6867.$$

The variance is $.6867 - (.8218)^2 = .01134$. So the posterior variance is

$$15.6027 + .01134(\overline{X}_i - 103.1842)^2. \tag{8.6}$$

The variances as well as 90% confidence intervals are also given in Table 8.4. The intervals were obtained by adding and subtracting 1.645 standard deviations from the posterior mean. This is in line with the Bayesian Central Limit Theorem. A plot of the posterior density of θ_1 revealed that the normal approximation was reasonable. A comparison with Table 8.1 reveals that nine of the ten intervals enclose the true value. It is equally interesting that in only four of the ten cases is the Bayes estimate closer to the true value than $\overline{X}_i$. On the other hand, the sum of the squared errors for the sample means is 166.68 while for the Bayes estimates it is 115.92. This is a fair comparison since both estimates are usually derived under squared error loss.

The next task is to see how well the Gauss-Hermite method works on these data. To give this method a full workout, no attempt will be made to reduce the dimension of the problem through a transformation. Proceeding directly from (7.48) the posterior distribution is

$$\pi^*_{22}(\sigma^2,\tau^2 \mid \boldsymbol{x}) \propto (\sigma^2)^{-20}(\sigma^2 + 5\tau^2)^{-4.5} exp\left(-\frac{1691.41}{\sigma^2} - \frac{1844.2575}{\sigma^2 + 5\tau^2} \right). \quad (8.7)$$

As suggested when this method was introduced, the log transformation is effective when the variables are variances. The transformation is $s = ln(\sigma^2)$, $t = ln(\tau^2)$. The Jacobian is $exp(s+t)$ and so the density is

$$\pi^*_{22}(s,t \mid \boldsymbol{x}) \propto e^{t-19s}(e^s + 5e^t)^{-4.5} exp\left(-\frac{1691.41}{e^s} - \frac{1844.2575}{e^s + 5c^t} \right). \quad (8.8)$$

Figures 8.1 through 8.4 contain plots of these two densities. Clearly the transformed variables are better approximated by a bivariate normal density.

In essentially no time at all the following mean vector and covariance matrix were obtained:

$$\boldsymbol{\mu} = \begin{bmatrix} 4.5152 \\ 4.6008 \end{bmatrix} \qquad \Sigma = \begin{bmatrix} 0.05392 & -0.01322 \\ -0.01322 & 0.48247 \end{bmatrix}.$$

These were then used to evaluate the four integrals needed to obtain the same posterior quantities obtained using adaptive Gaussian integration. The results displayed in Table 8.5 show that the Gauss-Hermite formula provided acceptable answers.

Figure 8.1
Posterior Density of Variances

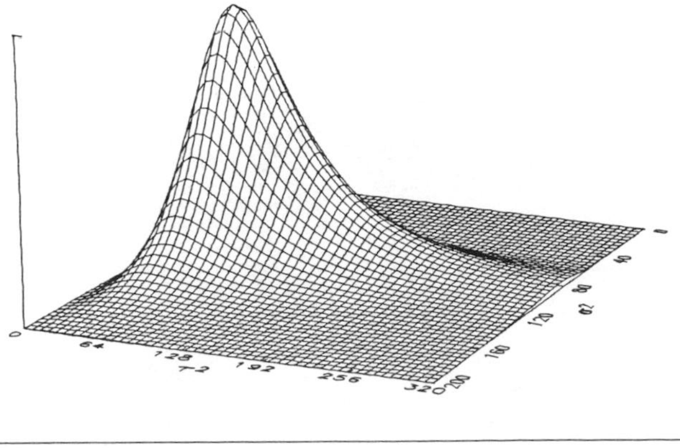

Figure 8.2
Posterior Density of Variances

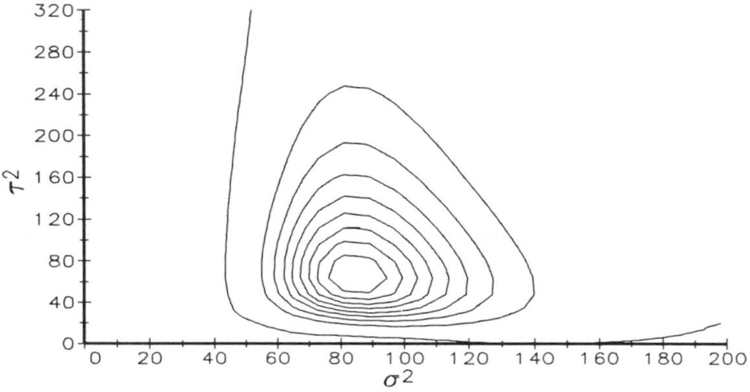

Figure 8.3
Posterior Density of log(variance)

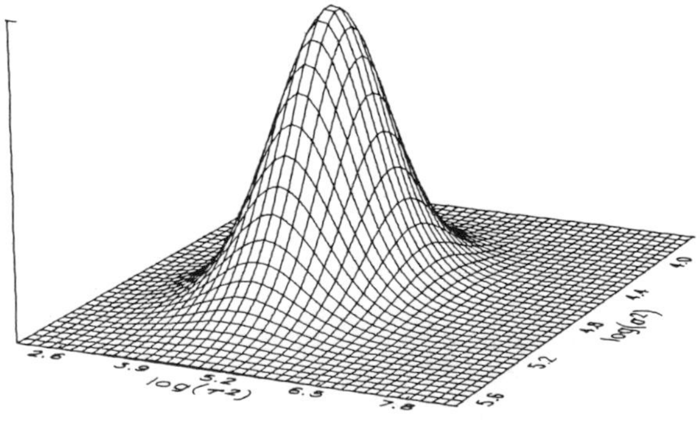

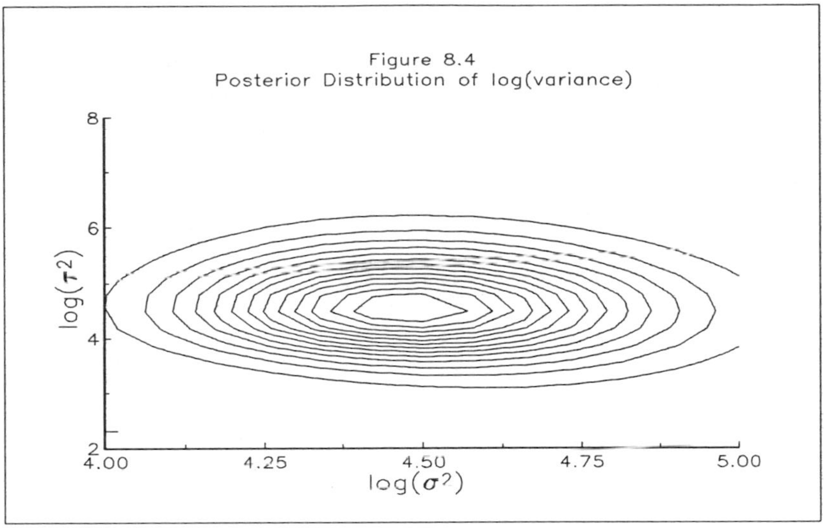

Table 8.5

Data Set 1 — Gauss-Hermite Integrals

Function	*integral*	*ratio to first integral*	*correct result*	*relative error*
1	101,370,491	—	—	—
$z = \dfrac{5e^t}{e^s + 5e^t}$	83,320,419	.8219	.8218	.00012
$z^2 = \left(\dfrac{5e^t}{e^s + 5e^t}\right)^2$	69,622,066	.6868	.6867	.00015
$v = \dfrac{e^s(e^s + 50e^t)}{50(e^s + 5e^t)}$	1,581,679,193	15.6030	15.6027	.00002

We next tried Monte Carlo integration on these data using the bivariate posterior with the log transformation (8.8). The mean and covariance matrix that were obtained from the Gauss-Hermite integration are used for the t-distribution. Nine degrees of freedom were used as that is the number of degrees of freedom associated with the empirical Bayes estimate of τ^2. A small sample of 1000 random observations was simulated to estimate the number of iterations needed. For a 50% probability of achieving three significant digits a sample size of 81,284 is required. This would have taken about 5.5 hours on my computer and so I settled for 10,000 observations. This would yield a median relative error of .0014. In general , given the sample size n the median number of significant digits is

$$.5log_{10}\left(\frac{n\mu^2}{1.8225\sigma^2}\right). \tag{8.9}$$

The results are in Table 8.6. The actual relative errors seem reasonably close to the predicted value.

Table 8.6

Data Set 1 — Monte Carlo Integrals

Function	*expected value*	*correct result*	*relative error*
$z = \frac{5e^t}{e^s + 5e^t}$	.8194	.8218	.00292
$z^2 = \left(\frac{5e^t}{e^s + 5e^t}\right)^2$	.6834	.6867	.00481
$v = \frac{e^s(e^s + 50e^t)}{50(e^s + 5e^t)}$	15.6362	15.6027	.00215

Table 8.7

Data Set 1 — Tierney-Kadane integrals

Function	*expected value*	*correct result*	*relative error*
$z = \frac{5e^t}{e^s + 5e^t}$	.8275	.8218	.00694
$z^2 = \left(\frac{5e^t}{e^s + 5e^t}\right)^2$	.6943	.6867	.01107
$v = \frac{e^s(e^s + 50e^t)}{50(e^s + 5e^t)}$	15.6652	15.6027	.00401

The Tierney-Kadane approach was next tried on Data Set 1 again using a logarithmic transformation of the variances. The results of the three basic integrals are given in Table 8.7. These values are not as good as those obtained by the other methods.

2. One-Way Model, Data Set 2

The next example uses Data Set 2. Its large size should imply time savings when the Gauss-Hermite formula is used along with an increase in the error. The data used as x_{ij} in (7.45)—(7.50) are Y_{ij}/P_{ij}, the relative frequencies for each class and year. Furthermore, only the first six of the seven years will be used. The results will then be used to forecast the number of claims in the seventh year. Letting θ_i be the true ratio for the ith class, the two levels of the model are

$$x_{ij} \mid \theta_i,\sigma^2 \sim N(\theta_i,\sigma^2/P_{ij}) \qquad \theta_i \mid \mu,\tau^2 \sim N(\mu,\tau^2) \tag{8.10}$$

If a realistic model is a Poisson distribution for the frequency counts, then the appropriate first level variance would be θ_i/P_{ij}. The formulas of Chapter 7 would not apply since the quantity of interest appears in the variance term. An analysis of the Poisson model is done in Chapter 9. The same techniques used for analyzing Data Set 1 apply here. The analog of (8.3) is

$$I_t(\delta) = \frac{\delta^{62.5}}{\left[\sum \frac{P_i}{P_i+\delta}\right]^{1/2} \left[\prod (P_i+\delta)\right]^{1/2}}$$

$$\times \left[62.235 + \sum \frac{P_i\delta}{2(P_i+\delta)}(\hat{\theta}_i - \hat{\mu})^2\right]^{-(381-t)} \Gamma(381-t)$$

$$\hat{\theta}_i = \sum_{j=1}^{6} P_{ij}X_{ij}/P_i, \qquad P_i = \sum_{j=1}^{6} P_{ij}, \qquad \hat{\mu} = \frac{\sum \frac{P_i}{P_i+\delta}\hat{\theta}_i}{\sum \frac{P_i}{P_i+\delta}}. \tag{8.11}$$

The four unmarked sums and the product are taken over the 130 (of 133) classes for which $P_i > 0$. In addition, in 13 of the years there was no exposure, so $N = 767$ total observations. To identify the scope of this function the mode was found (by the simplex method) to be $\delta = 109$ and the value of $I_0(109)/\Gamma(381)$ is $exp(-1824)$. To keep quantities in the range of the computer a constant 1800 was added to the log of $I_t(\delta)$ before exponentiation. Adaptive Gaussian quadrature was used for all the integrals that follow. The initial interval used was (0,50) and a convergence tolerance of 0.001 was used, producing approximately eight significant digits. Most of the computational time is used evaluating $I_t(\delta)$ and the fact that 130 integrals are being done simultaneously adds little to the effort. One round of integrals took about five minutes on an 8 MHz AT. Careful examination of the integrand will reveal that $\int_0^\infty \delta I_t(\delta)d\delta$ will not exist, although $\int_0^\infty \delta^{-1} I_t(\delta)d\delta$ will always exist for $k > 5$ (when the constant prior is used). The same restriction applies for the existence of

$E(\tau^2 \mid \boldsymbol{x}) = \int_0^\infty \delta^{-1} I_1(\delta)d\delta / 380 \int I_0(\delta)d\delta$. For

$$E(\sigma^2 \mid \boldsymbol{x}) = \int_0^\infty I_1(\delta)d\delta / 380 \int_0^\infty I_0(\delta)d\delta$$

to exist it is only necessary that $k > 3$. These results are established in Klugman (1987).

The first set of integrals was used to evaluate the posterior means of the variances. The corresponding Bühlmann-Straub empirical Bayes quantities were also found. They are:

$E(\sigma^2 \mid \boldsymbol{x}) = 0.19760$ $\qquad$ $\hat{\sigma}^2 = 0.19540$

$E(\tau^2 \mid \boldsymbol{x}) = 0.0017747$ $\qquad$ $\hat{\tau}^2 = 0.00062318$

$\delta^{BEB} = E(\sigma^2 \mid \boldsymbol{x})/E(\tau^2 \mid \boldsymbol{x}) = 111.34$ $\qquad$ $\delta^{EB} = \hat{\sigma}^2/\hat{\tau}^2 = 313.55$

The next set of integrals was used to obtain the posterior means of $\theta_1,\ldots,\theta_{133}$. Values for six selected classes are presented in Table 8.8 in the column headed *Bayes*. The first three classes used the smallest, median, and largest payroll while the second three classes had the smallest, median, and largest $\hat{\theta}_i$.

Also given in Table 8.8 are two other estimates. The one called *BayesEB* uses the ratio of the posterior means of σ^2 and τ^2 directly. Its advantage is that only two integrals are required. However, it is not a true Bayes estimate. The formula is

$$\frac{P_i}{P_i+111.34}\hat{\theta}_i + \frac{111.34}{P_i+111.34}(0.040480),$$

$$0.040480 = \hat{\mu} = \frac{\sum \frac{P_i}{P_i+111.34}\hat{\theta}_i}{\sum \frac{P_i}{P_i+111.34}}. \tag{8.12}$$

The last column, headed *EB*, contains the Bühlmann-Straub estimates. The formula is

$$\frac{P_i}{P_i+313.55}\hat{\theta}_i + \frac{313.55}{P_i+313.55}(0.039499),$$

$$0.039499 = \hat{\mu} = \frac{\sum \frac{P_i}{P_i+313.55}\hat{\theta}_i}{\sum \frac{P_i}{P_i+313.55}}. \tag{8.13}$$

Table 8.8

Data Set 2 — Frequency estimates, one way model

Class	P_i	$\hat{\theta}_i$	*Bayes*	*BayesEB*	*EB*
4	0.037	0.0	0.04045	0.04066	0.03949
11	1,053.126	0.04463	0.04422	0.04423	0.04345
112	93,383.540	0.00188	0.00193	0.00193	0.00201
70	287.911	0.0	0.01142	0.01129	0.02059
20	11,075.310	0.03142	0.03151	0.03151	0.03164
89	620.968	0.42997	0.36969	0.37075	0.29896

The first two estimates are similar in all cases, but the Bühlmann-Straub estimates are a bit different. This is due to the difference in the estimates of the second level variance, τ^2.

The Gauss-Hermite formula was used on the same data set. The six calculated posterior means given in Table 8.8 were each identical to five decimal places. The sum of the absolute differences of the 133 estimates was just 0.0000004. Finding the mean vector and covariance matrix and the 133 integrals took less than two minutes. Next, Monte Carlo integration was tried. A preliminary sample indicated that a sample size of 679 would yield two significant digits. The average absolute relative error of the 133 posterior means turned out to be 0.001, a better level of accuracy than was expected. The posterior means were also obtained with the Tierney-Kadane formula taking about two hours on my AT. The average absolute relative error was .00027. This is clearly an acceptable level of accuracy. A large number of simulations would be needed to match this with the Monte Carlo method.

The next step is to predict the number of claims in the seventh year. For year seven in class i the expected number of claims is $P_{i7}\theta_i$ while the variance is $P_{i7}\sigma^2$. The expected number of claims will be P_{i7} times the point estimate. In Table 8.9 the point estimates for the six classes used in Table 8.8 are presented along with the actual number of claims, Y_{i7}. At the bottom of each column a measure of prediction success is presented. It is

$$\sum_{i=1}^{133}(P_{i7}\theta_i^{est} - Y_{i7})^2/P_{i7} \tag{8.14}$$

where θ_i^{est} is the estimate of θ_i that is being evaluated. The Bayes estimates are seen to have been slightly superior at predicting the next value.

Table 8.9

Data Set 2 — Predicted frequencies for year seven

Class	P_{i7}	Y_{i7}	*Bayes*	*BayesEB*	*EB*	*StdDev*
4	0.0	0	0.0	0.0	0.0	0.0
11	229.83	8	10.16	10.16	9.99	7.37
112	18,809.67	45	36.34	36.34	37.82	66.82
70	54.81	0	0.63	0.63	1.13	3.51
20	1,315.37	22	41.45	41.45	41.62	17.04
89	79.63	40	29.44	29.42	23.81	4.23
Prediction error			13.20	13.21	15.55	

The other advantage of the Bayes approach is that posterior variances are available. From (7.42) we have

$$Var(Y_{i7} \mid \boldsymbol{x}) = P_{i7}E(\sigma^2 \mid \boldsymbol{x}) + P_{i7}^2 Var(\theta_i \mid \boldsymbol{x}) \tag{8.15}$$

and from (7.38)

$$Var(\theta_i \mid \boldsymbol{x}) = \int_0^\infty \frac{1}{P_i+\delta}\left[1 + \frac{\delta}{P_i+\delta}\frac{1}{\sum \frac{P_s}{P_s+\delta}}\right] I_1(\delta)d\delta \Big/ \int_0^\infty I_0(\delta)d\delta$$

$$+ \int_0^\infty \left[\frac{P_i}{P_i+\delta}\hat{\theta}_i + \frac{\delta}{P_i+\delta}\hat{\mu}\right]^2 I_0(\delta)d\delta \Big/ \int_0^\infty I_0(\delta)d\delta - \left[E(\theta_i \mid \boldsymbol{x})\right]^2. \tag{8.16}$$

The two numerator integrals in (8.16) are no problem. The standard deviations are presented in Table 8.9. In addition, 90% prediction intervals were constructed by adding and subtracting 1.645 standard deviations from the point estimate. Of the 128 classes that had exposure in the seventh year, 124 of the prediction intervals enclosed the true value. Either the normal approximation to the posterior density is not too good, or the normal model for frequencies is not too good, or the one-way model is not too good, or the sample is not too good. These questions will be addressed in later sections.

3. Empirical Bayes Style Approaches

This approach was illustrated in the previous section with Data Set 2. If the posterior mean is the quantity of interest then the only relevant quantity is $\delta = \sigma^2/\tau^2$. Using the Gauss-Hermite results this becomes $\delta = e^s/e^t$. Substituting the expected values of s and t (-1.6231 and -6.3465 respectively) yields $\delta = 112.55$. A new value of $\hat{\mu}$ is also needed. Doing as in (8.12) and (8.13) gives $\hat{\mu} = .040472$. For example, consider Class 20 (see Table 8.8). The estimate of the posterior mean is

$$\begin{aligned}\theta_{20}^* &= \frac{11{,}075.31}{11{,}075.31 + 112.55}(.03142) + \frac{112.55}{11{,}075.31 + 112.55}(.040472) \\ &= .03151.\end{aligned}$$

Suppose it is desired to find the posterior variance of the next observation, as was done in (8.15). It is easy to see that all the terms are immediately available except for $Var(\theta_i \mid \boldsymbol{x})$. Recall that this can be written as $E(v_{ii}^* \mid \boldsymbol{x}) + Var(\theta_i^* \mid \boldsymbol{x})$. Since v_{ii}^* depends only on the elements of F and

G the same substitution can be used here. It is the second term that is not easy to handle. For the one-way model, use (7.45) to obtain for Class 20

$$v^*_{20,20} = \frac{\sigma^2}{P_{20} + \sigma^2/\tau^2} + \left[\frac{\sigma^2/\tau^2}{P_{20} + \sigma^2/\tau^2}\right]^2 \Big/ \sum \frac{P_i}{\sigma^2 + P_i\tau^2} \tag{8.17}$$

and substituting the posterior means of $\sigma^2 = e^{-1.6231} = .19729$ and $\tau^2 = e^{-6.3465} = .0017529$ produces a value of .000017799. The final expression for the variance is

$$1315.37(.19729) + (1315.37)^2[.000017799 + Var(\theta_i^* \mid \boldsymbol{x})]$$

$$= 290.31 + (1315.37)^2 Var(\theta_i^* \mid \boldsymbol{x}). \tag{8.18}$$

When compared to the true variance of 290.36 we see that taking $Var(\theta_i^* \mid \boldsymbol{X})$ to be zero is not unreasonable.

The variance can be approximated using the empirical Bayes techniques from Chapter 3. In general we can write $\theta_i^* = g(\boldsymbol{\sigma})$ where $\boldsymbol{\sigma}$ is a vector containing the variance terms from the hierarchical model. We then have (using (3.34))

$$Var[g(\boldsymbol{\sigma}) \mid \mathbf{x}] \doteq (\nabla g)' Var(\boldsymbol{\sigma} \mid \boldsymbol{x})(\nabla g) \tag{8.19}$$

where ∇g is the gradient vector (the vector of first partial derivatives) and $Var(\boldsymbol{\sigma} \mid \boldsymbol{x})$ is the conditional covariance matrix of $\boldsymbol{\sigma}$.

For the one-way model with the natural log transform the vector is $\boldsymbol{\sigma} = (s,t)'$ and the function in question is

$$g(s,t) = \frac{P_i e^t}{e^s + P_i e^t}\hat{\theta}_i + \frac{e^s}{e^s + P_i e^t}\hat{\mu} = \frac{P_i e^t}{e^s + P_i e^t}(\hat{\theta}_i - \hat{\mu}) + \hat{\mu}. \tag{8.20}$$

It should be noted that $\hat{\mu}$ depends on s and t. To make the analysis tractable $\hat{\mu}$ will be taken as fixed. The partial derivatives are

$$-\frac{\partial g(s,t)}{\partial s} = \frac{\partial g(s,t)}{\partial t} = \frac{P_i e^s e^t}{(e^s + P_i e^t)^2}(\hat{\theta}_i - \hat{\mu}) = z_i(1 - z_i)(\hat{\theta}_i - \hat{\mu})$$

and so

$$Var(\theta_i^* \mid \boldsymbol{x}) \doteq (\hat{\theta}_i - \hat{\mu})^2 (z_i)^2 (1 - z_i)^2$$

$$\times [Var(s \mid \boldsymbol{x}) + Var(t \mid \boldsymbol{x}) - 2Cov(s,t \mid \boldsymbol{x})]. \quad (8.21)$$

For Class 20 in Data Set 2 the value is

$$Var(\theta_{20}^* \mid \boldsymbol{x}) \doteq (.03142 - .040472)^2 (.98994)^2 (.01006)^2$$

$$\times [.00318 + .02473 - 2(-.00085)] = 2.406(10)^{-10}.$$

When multiplied by $(1315.37)^2$ (see (8.18)) the additional factor is .00042 and we see that in this particular instance the extra work did not change the answer by much. One interesting point about this factor is the inclusion of the term $(\hat{\theta}_i - \hat{\mu})^2$. This reminds us that, as in regression analysis, it is more difficult to accurately predict values that are far from the center.

4. An Iterative Approach

Another approximation method that is available for the *HNLM* is not unlike the E-M algorithm that was designed for finding maximum likelihood estimates in the presence of missing data. Here $\boldsymbol{\theta}$ plays the role of the missing data while F and G are the parameters of interest. The expectation step has already been performed. It is the same one that was used above, namely given estimates of F and G estimate $\boldsymbol{\theta}$ with $E(\boldsymbol{\theta} \mid \boldsymbol{x}, F, G) = \boldsymbol{\theta}^*$. The approach to use for estimating the variance terms is to find the values that maximize the density $\pi(\boldsymbol{\theta}, F, G \mid \boldsymbol{x})$ where $\boldsymbol{\theta}$ is held fixed at the value $\boldsymbol{\theta}^*$. This latter density was obtained in general in (6.38) and for the *HNLM* in (7.44). It is just the product of the two densities used in Section 1 above. This necessarily leads to an iterative scheme since one result is needed to obtain the other. A reasonable place to start the iteration is with $\boldsymbol{\theta}^* = \hat{\boldsymbol{\theta}}$.

For example, again consider Class 20 in Data Set 2. From (7.44) the expectation step is

$$\boldsymbol{\theta}_{20}^* = \frac{11{,}075\tau^2}{\sigma^2 + 11{,}075\tau^2}(.03142) + \frac{\sigma^2}{\sigma^2 + 11{,}075\tau^2}\hat{\mu}, \quad \hat{\mu} = \frac{\sum \dfrac{P_i \hat{\theta}_i}{\sigma^2 + P_i \tau^2}}{\sum \dfrac{P_i}{\sigma^2 + P_i \tau^2}}. \quad (8.22)$$

The maximization step uses the density that is given in (7.44). For the one-way model it reduces to

$$\pi^*(\boldsymbol{\theta}^*,\sigma^2,\tau^2 \mid \boldsymbol{x}) \propto (\sigma^2)^{-N/2}(\tau^2)^{-(k-1)/2} exp[-WSS/2\sigma^2$$
$$- \sum P_i(\theta_i^* - \hat{\theta}_i)^2/2\sigma^2 - \sum \theta_i^{*2}/2\tau^2 + (\sum \theta_i^*)^2/2k\tau^2]. \qquad (8.23)$$

Taking twice the negative natural logarithm of this density produces the following function:

$$Nln(\sigma^2) + (\mathrm{k}-1)ln(\tau^2) + WSS/\sigma^2 + \sum P_i(\theta_i^* - \hat{\theta}_i)^2/\sigma^2$$
$$+ \sum \theta_i^{*2}/\tau^2 - (\sum \theta_i^*)^2/k\tau^2. \qquad (8.24)$$

Setting the two partial derivatives equal to zero produces the following maximizing values:

$$\sigma^2 = \frac{WSS + \sum P_i(\theta_i^* - \hat{\theta}_i)^2}{N} \qquad \tau^2 = \frac{\sum \theta_i^{*2} - (\sum \theta_i^*)^2/k}{k-1}. \qquad (8.25)$$

Careful inspection will reveal that these are just the within and between mean squares computed with respect to the estimates θ_i^*. It should also be clear that these are pseudo-estimators in the sense mentioned in Chapter 5 as in both formulas σ^2 and τ^2 appear on the right hand side. In the usual derivation of pseudo-estimators they are created by looking at promising sums of squares and then adjusting them to be unbiased when expectations are taken conditioned on the parameters on the right hand side. The above development is more natural since the sums of squares came directly from the density function. Unlike the Bühlmann-Straub estimates, these cannot yield negative values for the variances. However, it is possible for τ^2 to be zero.

The iterations were performed on Data Set 2 starting with $\hat{\boldsymbol{\theta}}$. The results are given in Table 8.10. Although only the estimates of θ_{20} are displayed, it is clear from (8.25) that all 133 values must be obtained at each iteration. The result appears to be closer to the Bayes results than to the Bühlmann-Straub answer.

Table 8.10

Data Set 2 — EM Estimates

Iteration	σ^2	τ^2	θ^*_{20}
1	.16228	.0028555	.031470
2	.16492	.0016929	.031501
3	.16774	.0014725	.031513
4	.16899	.0014019	.031518
5	.16950	.0013758	.031520
6	.16971	.0013655	.031521
7	.16980	.0013614	.031521
8	.16983	.0013598	.031521
9	.16985	.0013591	.031521
10	.16985	.0013589	.031521
11	.16986	.0013588	.031521
12	.16986	.0013587	.031521
13	.16986	.0013587	.031521

5. Other Priors

Data Sets 1 and 2 will now be analyzed with the other priors mentioned at the end of Chapter 7. For the one-way model the first alternative mentioned was

$$\pi(\sigma^2,\tau^2) \propto (\sigma^2)^{-1}[\prod(\sigma^2 + P_i\tau^2)]^{-1/k}. \tag{8.26}$$

Call this Prior 2 for later reference. Another suggestion involved the geometric mean of the diagonal elements of $AGA' + F$. It is

$$[\prod(\sigma^2/P_{ij} + \tau^2)]^{1/N}.$$

Raising the reciprocal to the second power as suggested (and rearranging the constants) yields

$$\pi(\sigma^2,\tau^2) \propto [\prod(\sigma^2/P_{ij} + \tau^2)]^{-2/N} \tag{8.27}$$

Call this Prior 3 and let Prior 1 refer to the constant function. Prior 3 will be a little more time consuming to apply since the product is over all N observations and not just over the k groups.

If the problem is to be solved by Gaussian integration the transformation (7.49) needs to be done. For both priors the density in (7.50) will still be amenable to analytic integration with respect to α. The integrations with these priors were performed on Data Set 1 and the results are given in Table 8.11. It can be seen that the choice of prior does make a bit of difference, but not an enormous amount.

Table 8.11

Data Set 1 — Posterior Expected Values, Three Priors

Function	*Prior 1*	*Prior 2*	*Prior 3*
$z = \dfrac{5e^t}{e^s + 5e^t}$	.8275	.7830	.7776
$z^2 = \left(\dfrac{5e^t}{e^s + 5e^t}\right)^2$	.6943	.6266	.6170
$v = \dfrac{e^s(e^s + 50e^t)}{50(e^s + 5e^t)}$	15.6652	14.1337	14.1311

Prior 2 was also tried on Data Set 2 with Gauss-Hermite integration. Using Prior 2 added no extra computation time (as would happen for Data Set 1, but not for Prior 3 with Data Set 2). Results for six of the groups are presented in Table 8.12. In two of the classes (70 and 89) the change in prior made a difference in the result. What they have in common is being a long way from the middle ($\hat{\mu}$) and having a moderate exposure. But these are the only interesting cases. When the exposure is large or small the credibility factor will be one or zero, regardless of how the problem is approached. When the sample mean is near the middle the use of shrinking will have little effect on the answer. On the other hand, it should be noted that in the two cases where the answers differed, the change is only one unit in the third significant digit.

Table 8.12

Data Set 2 — Posterior Means, Two Priors

Class	P_i	$\hat{\theta}_i$	*Prior 1*	*Prior 2*
4	0.037	0.0	0.04045	0.04044
11	1,053.126	0.04463	0.04422	0.04421
112	93,383.540	0.00188	0.00193	0.00193
70	287.911	0.0	0.01142	0.01155
20	11,075.310	0.03142	0.03151	0.03151
89	620.968	0.42997	0.36969	0.36887

6. Diagnostics

In this Section the assumptions of normality and constant variance in the one-way model will be checked for Data Set 3. Adaptive Gaussian integration was used to obtain

$$\hat{\sigma}^2 = E(\sigma^2 \mid \boldsymbol{x}) = 0.00023537 \text{ and } \hat{\tau}^2 = E(\tau^2 \mid \boldsymbol{x}) = 0.00011470.$$

The ratio $\hat{\delta} = 2.0521$ was used to obtain $\hat{\mu}$. Four sets of residuals were then obtained. The first three are relatively simple:

$$\textit{Level 1}: \quad r_{ij} = \frac{X_{ij} - \hat{\theta}_i}{\hat{\sigma}/\sqrt{P_{ij}}} \tag{8.28}$$

$$\textit{Level 2}: \quad r_i = \frac{\hat{\theta}_i - \hat{\mu}}{\sqrt{\hat{\sigma}^2/P_i + \hat{\tau}^2}} \tag{8.29}$$

$$\textit{Overall1}: \quad r_{ij} = \frac{X_{ij} - \hat{\mu}}{\sqrt{\hat{\sigma}^2/P_i + \hat{\tau}^2}} \tag{8.30}$$

The first two come directly from (7.77) and (7.78). For the one-way model each of these distributions has a diagonal covariance matrix and so the standardization is easy.[18] The third residual comes from (7.79) and does the standardization by dividing by the square root of the variance.

[18] This does not mean that the estimated residuals are independent. These estimates depend on the variance estimates which in turn depend on all the data.

However, a careful look at (7.79) reveals that the covariance matrix is only block diagonal. The ith block is $n_i \times n_i$ and has $\sigma^2/P_i + \tau^2$ on the diagonal and τ^2 off the diagonal. A true standardization must be done using the square root of the inverse of this matrix. Residuals calculated this way (the Choleski factorization was used) are called *Overall2*. So there are two ways of getting residuals that compare the observations to the grand mean.

The first investigation was done to see if there is any relationship between the variance and either the pure premium or the exposure. In Figure 8.5 the Level 2 residuals are plotted against the exposures P_i. There is clearly no linear relationship and it does not appear that the variance is changing with the exposure. We should expect a greater range where the exposures are small just because there are so many more observations there. To make the graph more readable the five classes with the highest exposure were removed. The exposures were 543, 830, 1075, 1249, and 2786 while the residuals were −0.92, −0.55, −1.49, −1.57, and −1.58. This is a bit suspicious, as the residuals associated with the nine largest exposures are all negative. As this is workers' compensation data this means that occupations with the largest work force tend to be the ones with low claim levels. This fact does not invalidate the model.

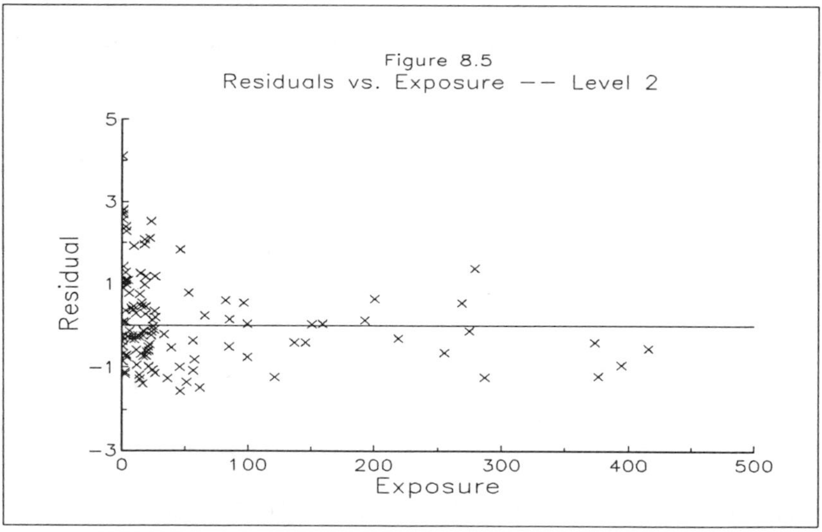

The next plot (Figure 8.6) is the only one that can compare residuals to pure premiums. This is done for the level 1 residuals, plotting them against the group means, $\hat{\theta}_i$. There are only 724 (not $6 \times 121 = 726$) values since two of the years had no exposure. The twelve largest means were not plotted (6 at $\hat{\theta} = 0.08928$ and 6 at $\hat{\theta} = 0.11507$). There was no pattern here as the values ranged from −1.76 to 3.97. These variances are surprisingly stable as the means increase. It would be expected that the variance would be small where the mean is small as there is very little room to move in the negative direction and a small positive movement represents a large increase in the pure premium. Of more interest are the four residuals that are between 4 and 6. A complete investigation would look at them to see if there is a logical explanation for these values. It is possible that the analysis would be greatly improved if these values were left out. This is done later in this Section.

Figure 8.6
Residuals vs. Group Means -- Level 1

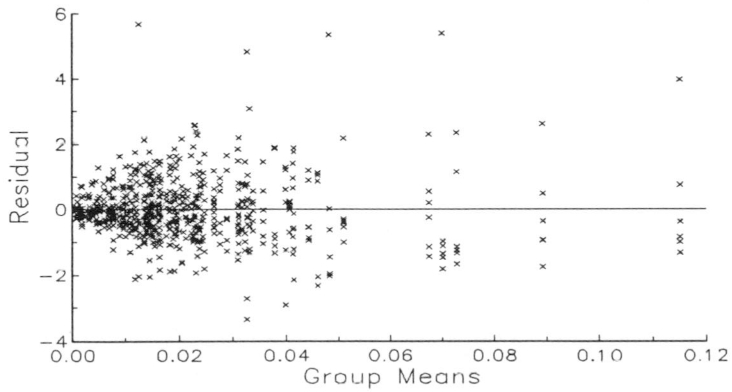

The next three plots compare the data to the exposures. The values with the six highest exposures (342, 378, 428, 487, 556, 594) were not plotted in each case. In Figure 8.7 the level 1 residuals are plotted against the individual exposures. The six residuals that were left off range from −0.29 to 0.42. The plot seems to indicate that there is a slight decrease in variance as the exposure increases. The second plot (Figure 8.8) shows the overall residuals, version 1, against the exposures. They look terrible, but this is most certainly because (8.29) used an inappropriate shortcut. The proper residuals appear in Figure 8.9. The six missing values range from −1.57 to −.39. This is the same phenomenon that was observed with level 2. Again there appears to be a decreasing trend in the variance. This contradicts the findings of Hewitt (1967). He found that the variance did not decrease as fast as being inversely proportional to exposure. Standardizing by using the exposure should produce residuals that increase in variance as the exposure increases. This indicates that the parameter uncertainty model is not appropriate for these data. A formal test of this will be conducted later in the Chapter.

Figure 8.7
Residuals vs. Exposures — Level 1

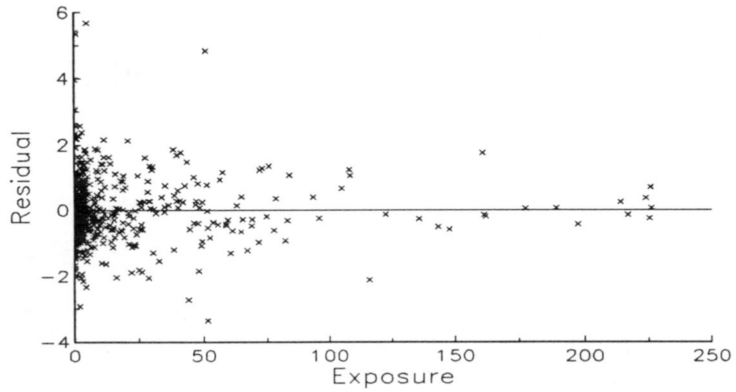

Figure 8.8
Residuals vs. Exposures — Overall1

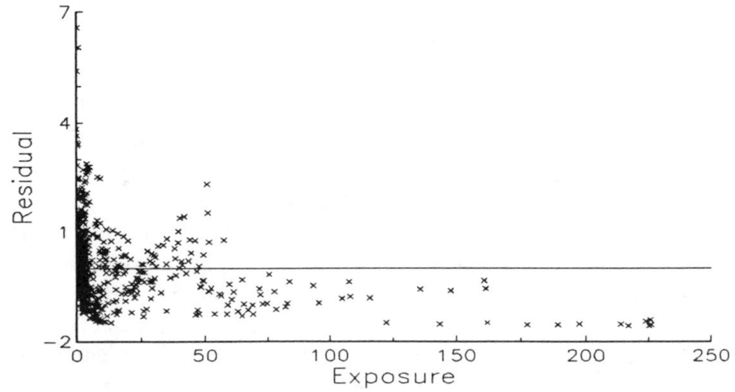

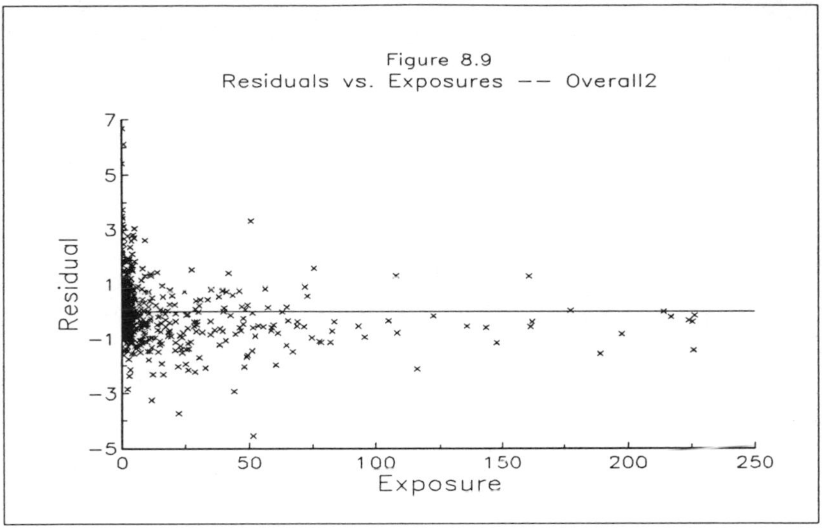
Figure 8.9
Residuals vs. Exposures -- Overall2
Residual
Exposure
7
5
3
1
−1
−3
−5
0
50
100
150
200
250

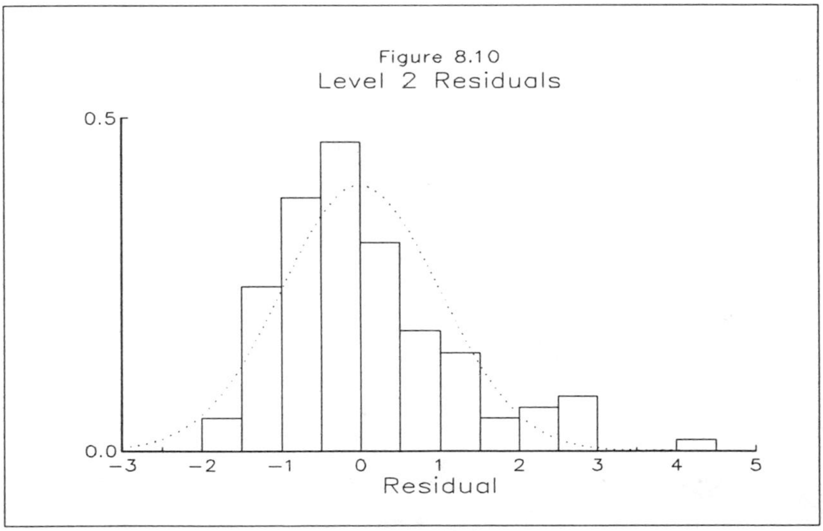
Figure 8.10
Level 2 Residuals
0.5
0.0
−3
−2
−1
0
1
2
3
4
5
Residual

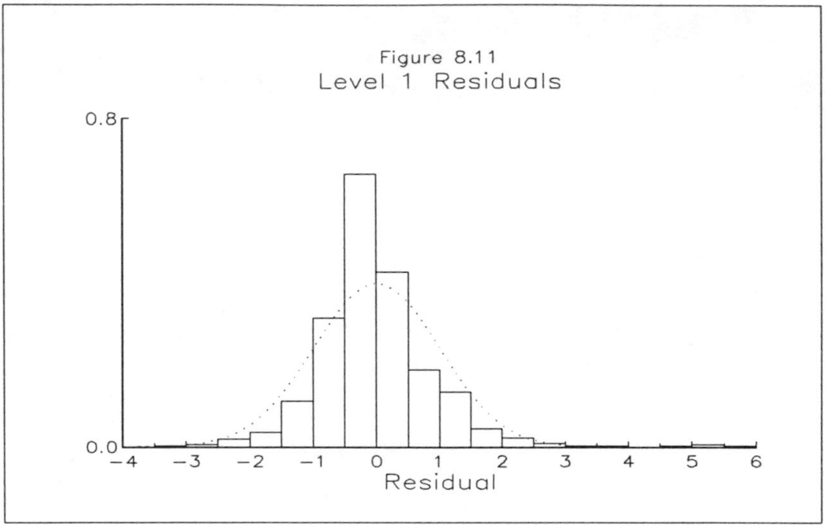

Figure 8.11
Level 1 Residuals

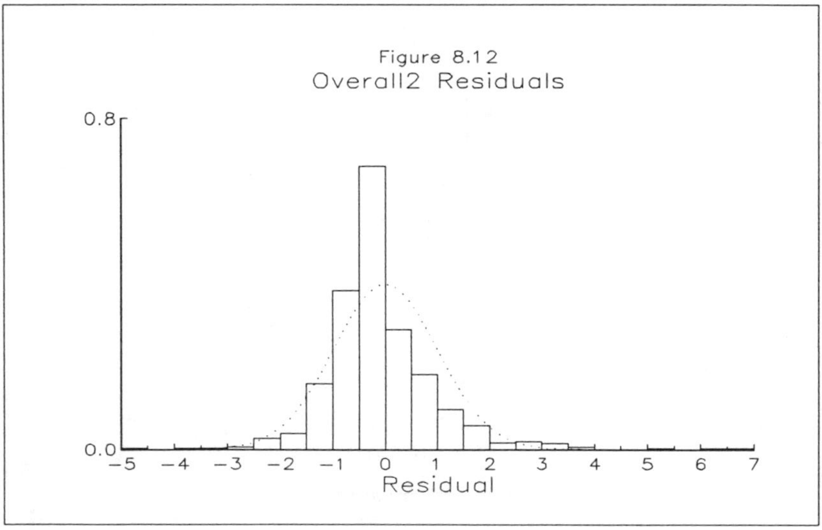

Figure 8.12
Overall2 Residuals

The next item to investigate is the assumption of normality. Histograms of the three sets of residuals appear in Figures 8.10 through 8.12 along with the standard normal *pdf.* All of them look at least somewhat normal although there does seem to be a slight bias towards negative values and some skewness to the right. Had the skewness been more pronounced, a lognormal distribution might be worth considering as an alternative model. This model is discussed in the next Chapter.

Table 8.13

Data Set 3 — Level 1 Outliers

Group	*Residual*	*Pure Premiums*					
44	4.82	.032	.026	.026	.034	.043*	.033
75	5.35	.016	.011	.145*	.037	.048	.003
85	5.40	.261*	.000	.014	.016	.000	.000
57	5.68	.005	.011	.011	.009	.018	.053*

**outlying value*

Return now to the four outliers that were observed in Figure 8.6. In Table 8.13 the values for these four classes are presented in more detail. The values from groups 75 and 85 are especially bothersome. To see what might happen, all 24 items were removed from the sample. The estimates became (again using adaptive Gaussian integration) $\hat{\sigma}^2 = 0.00019061$, $\hat{\tau}^2 = 0.00012496$, and $\hat{\delta} = 1.5253$. This is a significant change in the estimate of σ^2. This change did not have much of an effect on the residual plots (other than having the four outlying points removed) and in particular had very little effect on the level 2 residuals as seen in Figure 8.13. The new level 1 residuals are plotted in Figure 8.14 and do appear to be a little bit closer to the normal curve than those in Figure 8.11.

It is worth repeating that in a serious investigation it would be extremely important to check out these four groups to determine if there was a particular cause for the unusual results (e.g., coding errors, a freak accident that produced numerous claims).

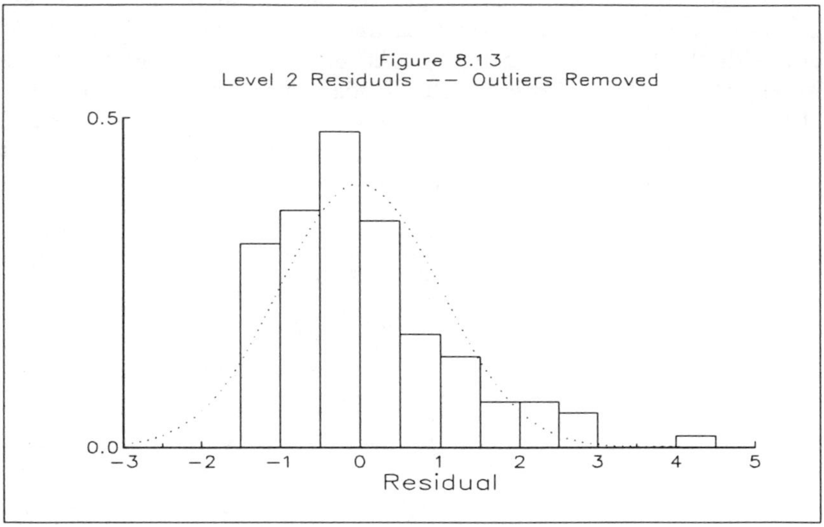
Figure 8.13
Level 2 Residuals -- Outliers Removed
0.5
0.0
−3
−2
−1
0
1
2
3
4
5
Residual

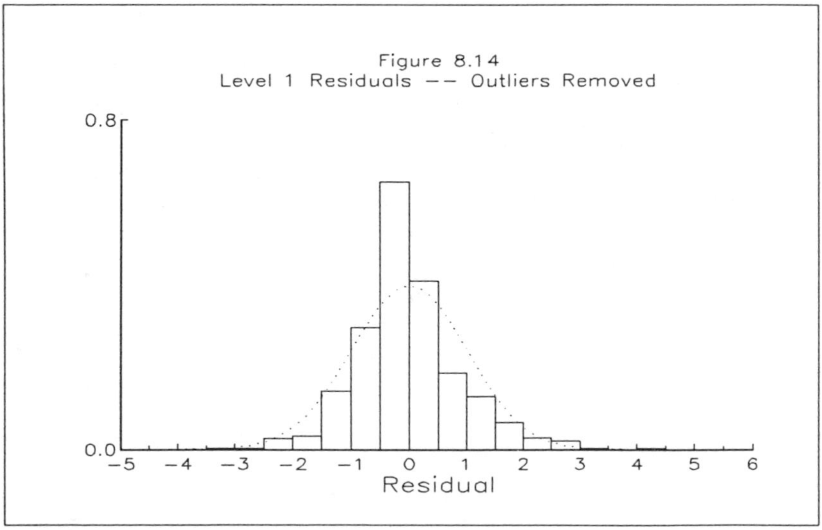
Figure 8.14
Level 1 Residuals -- Outliers Removed
0.8
0.0
−5
−4
−3
−2
−1
0
1
2
3
4
5
6
Residual

7. Two-way Model, Data Set 4

The simple version ($\gamma^2 = 0$) was used along with the version II formulas to analyze this data set. Formulas (7.53) and (7.54) were used to obtain the posterior density of σ^2, τ_α^2, and τ_β^2. The prior density $\pi_{22} \propto 1$ was used. Logarithms of these values were substituted to make finding the

Table 8.14

State and Occupation Estimates

State	*EB*	*G-H*	*Occ.*	*EB*	*G-H*
1	.009490	.009581	1	.071087	.070892
2	.049227	.049304	2	.079194	.078956
3	.101277	.099846	3	.020659	.020612
4	.031014	.031094	4	−.026646	−.025719
5	.006812	.006916	5	.053947	.053859
6	.032216	.032324	6	.187278	.187143
7	.009950	.010114	7	.041018	.040960
8	.111040	.111037	8	.092330	.091805
9	.006533	.006755	9	.053446	.053180
10	.052620	.052626	10	.012075	.012031
			11	.036054	.035973
			12	.020628	.020581
			13	.030890	.030818
			14	.039358	.039257
			15	.051675	.051548
			16	.081576	.081473
			17	.030695	.030631
			18	.022275	.022305
			19	.017960	.018116
			20	.025720	.025642
			21	.014436	.014392
			22	.030774	.030696
			23	.003541	.003514
			24	.035770	.035703
			25	−.000294	−.000375

maximum easier as well as to improve the accuracy of Gauss-Hermite integration. The four point integration formula produced posterior means of .24758, .0022785, and .0017099 respectively. These were then inserted into the version II formula for $\boldsymbol{\theta}^*$ to obtain empirical Bayes estimates. As well, Gauss-Hermite integration was performed to obtain the posterior mean of $\boldsymbol{\theta}^*$. Both sets of estimates are displayed in Table 8.14. It is clear there is little difference in the two sets of estimates. The negative values for some of the estimates should not be disturbing because the only relevant quantity is the sum of individual pairs of state and occupation factors. Recall that the results will be unchanged if a constant is added to each state factor and then subtracted from each occupation factor.

Estimates of the expected value for each state/occupation combination are then obtained by adding the relevant factors. As well, estimates based on the exposure-weighted average $(\overline{x}_{ij} = \sum_t P_{ijt}x_{ijt}/\sum_t P_{ijt})$ are available (recall that $\hat{\boldsymbol{\theta}}$ cannot be computed with the version II formulas). These, along with the experience from year 7 (which was held out of the sample), are displayed for ten selected combinations in Table 8.15. The measure of prediction error in (8.14) was also evaluated for each estimator.

Table 8.15

Predicted and Actual Values for Year 7

State	*Occ.*	$\overline{x}_{ij}$	*EB*	*G-H*	x_{ij7}
1	1	.07593	.08058	.08047	.10929
1	2	.02587	.08868	.08854	.10172
1	3	.02517	.03015	.03019	.02777
1	4	.00000	−.01726	−.01614	.00000
1	5	.06479	.06344	.06344	.07869
2	1	.07611	.12031	.12020	.33565
2	2	.11306	.12842	.12826	.09343
2	3	.07363	.06989	.06992	.10437
2	4	.00000	.02258	.02359	.00000
2	5	.08823	.10317	.10316	.11444
Prediction error		43.76	50.69	50.67	

The negative values here are more disturbing but not surprising. A normal distribution was assumed and when the additive restriction is applied negative values are in the parameter space. Possible solutions are either to calculate the posterior means of $\alpha_i + \beta_j$ integrating only over the first quadrant (Monte-Carlo integration would have to be used) or to consider a transformation that recognizes that that distribution of loss counts is almost certainly skewed. The latter suggestion is discussed in Chapter 9.

The measures of prediction error indicate that this procedure did not add much value. It appears that the additive model does not accurately describe the relationship between state and occupation. One solution is to return to a one-way model, treating all 250 state-occupation cells as separate classes. We would expect this to improve upon using $\bar{x}_{ij}$ as an estimator due to the superiority of credibility estimators. A compromise between these approaches can be made by using the first two-way model discussed in Chapter 7. It allows each class to deviate toward $\bar{x}_{ij}$ from the strict additive relationship.

8. Linear Trend Model, Data Set 3

Data Set 3 was evaluated in Section 6 using the one-way model. Here we consider the same data with the linear trend model. It is a bit difficult to work with the posterior density in (7.70) both for maximization and integration as the support for each variable is not the real line. To remedy this problem, the following transformations were used:

$$s = ln\sigma^2 \quad t = ln\tau_\alpha^2 \quad u = ln\tau_\beta^2 \quad r = \rho/(1-\rho^2) \quad \rho = \tau_{\alpha\beta}/\sqrt{\tau_\alpha^2\tau_\beta^2}. \tag{8.31}$$

Using the simple prior density $\pi_{22} \propto 1$ the posterior mode is $\hat{s} = -8.4167$, $\hat{t} = -9.2884$, $\hat{u} = -14.798$, and $\hat{r} = .8398$. This translates to $\hat{\sigma}^2 = .0002211$, $\hat{\tau}_\alpha^2 = .00009249$, $\hat{\tau}_\beta^2 = .0000003743$, and $\hat{\tau}_{\alpha\beta} = .00000345$. Values of $\tilde{\alpha}$ and $\tilde{\beta}$ are given for the first five classes in Table 8.16.

Table 8.16

Linear Trend Model — Intercept and Slope Estimates

Class	$\tilde{\alpha}$	$\tilde{\beta}$
1	.02560	.00143
2	.01772	.00096
3	.00989	.00063
4	.00883	.00047
5	.01263	.00071

The next task is to see if the linear trend model is an improvement over the one-way model. Using the parameter estimates from Section 6 the SBC for the one-way model is 2871.67. The penalty is based on $p = 3$ and $n = 724$. For the linear trend model the SBC is 2889.67. Even with an additional penalty of 9.49 ($p = 5$ here) the linear trend model still rates as being superior.

9. Kalman Filter, Data Set 3

Two specific forms of the Kalman filter model were introduced in Chapter 7. One is an autoregressive model that allows some carryover from year to year. The model is set out in (7.12) and (7.13) and will be called model 1. The second model introduces linear growth and is set out in (7.14) and (7.15). It will be called model 2. Both models will be applied to Data Set 3. Again the last year of data will be held back to check the ability of each model to predict future observations.

The first step in the analysis is to apply the Kalman filter separately to each of the 121 classes. The parameters are estimated by maximizing the likelihood function (4.27). To have the maximization be over all real numbers the same transformations that were used for the linear trend model will be used here. For model 1 they are:

$$s = ln\sigma^2 \quad t = ln\gamma^2 \quad r = \rho/(1-\rho^2). \tag{8.32}$$

The maximizing value of t was so small (< -20) that $\gamma^2 = 0$ was used. The other values were $\hat{s} = -7.664$ and $\hat{r} = -.2672$ corresponding to $\hat{\sigma}^2 = .0004694$ and $\hat{\rho} = -.2504$. The negative value of $\hat{\rho}$ was surprising as it indicates an oscillating pattern. The model was re-run with $\rho = 0$ (which reduces to the standard model with an unchanging mean) and this decreased the loglikelihood by only 3.76. The SBC penalty (based on 482 observations, recalling that the first two observations in each class must be used to initialize the process) is 4.34 and so it is not likely that model will be much of an improvement.

For model 2 the parametrization is

$$s = ln\sigma^2 \quad t = ln\gamma_1^2 \quad u = ln\gamma_2^2 \tag{8.33}$$

and the estimates are $\hat{s} = -7.7602$, $\hat{t} = -19.0221$, and $\hat{u} = -8.5956$ corresponding to $\hat{\sigma}^2 = .0004264$, $\hat{\gamma}_1^2 = .000000005480$, and $\hat{\gamma}_2^2 = .0001849$. This model can be reduced to the linear model by setting $\gamma_1^2 = \gamma_2^2 = 0$. The likelihood decreases by 174.92 indicating that this extra variability belongs in the model.

The next step is to use the results of these Kalman filter calculations in an *HNLM* as outlined in (7.72)—(7.74). Again a transformation is required. It is

$$t = \ln g_1^2 \quad u = \ln g_2^2 \quad v = \nu/(1-\nu^2) \quad \nu = g_{12}/\sqrt{g_1^2 g_2^2}. \tag{8.34}$$

For both models the posterior mode of v turned out to be extremely close to zero and so $g_{12} = 0$ was used in all subsequent calculations. The posterior means of $exp(t)$ and $exp(u)$ were then found by Gauss-Hermite integration. For model 1 the posterior means are .00003975 for both variances while for model 2 they are .00004870 and .000004737 for g_1^2 and g_2^2 respectively.

The posterior means were then used to obtain empirical Bayes estimates of the parameters. The Kalman filter and credibility estimates for the first five classes for model 1 are given in Table 8.17. They are followed in Table 8.18 with the values for model 2. The agreement in the two parameter values for model 1 provides further indication that the autoregressive model adds little to this analysis. The credibility step makes some major adjustments, indicating that this step does have some value. For model two, the surprising result is that after the credibility step the slopes are all negative. There is considerable shrinkage here to the average slope, which is also negative.

Table 8.17

Parameter Estimates — Model 1

	Kalman filter		*Credibility*	
Class	$\hat{\theta}_{i,6}$	$\hat{\eta}_i$	$\tilde{\theta}_{i,6}$	$\tilde{\eta}_i$
1	.03263	.03263	.02181	.02182
2	.02191	.02191	.01729	.01729
3	.01174	.01174	.01332	.01332
4	.00999	.00999	.01363	.01362
5	.01291	.01293	.01506	.01507

The estimates were then used in (4.24) to predict the seventh year. For model 1 the predicted value is $-.2504\tilde{\theta}_{i,6} + 1.2504\tilde{\eta}_i$ while for model 2 it is $\tilde{\theta}_{i,6} + \tilde{\eta}_{i,6}$. The quality of these predictions along with those using the one-way model and the linear trend model are evaluated using (8.14). The results appear in Table 8.19. There is not much to choose from with the exception of the linear trend model, which was clearly inferior. This contradicts the finding that this model was superior by the SBC.

Table 8.18

Parameter Estimates — Model 2

	Kalman filter		*Credibility*	
Class	$\hat{\theta}_{i,6}$	$\hat{\eta}_{i,6}$	$\tilde{\theta}_{i,6}$	$\tilde{\eta}_{i,6}$
1	.03922	.00318	.02751	−.00057
2	.02035	.00117	.01856	−.00048
3	.01707	.00688	.01655	−.00034
4	.00449	−.00287	.01116	−.00046
5	.01459	.00461	.01596	−.00039

Table 8.19

Evaluation of Predictions

Model	*Model 1*	*One-way*	*Model 2*	*Linear Trend*
Prediction Error	.04075	.04134	.04673	.07576

10. Graduation

The graduation problem introduced in Chapter 4 will be analyzed using the formulas developed in Chapter 7. The major difference is that the smoothing parameter τ^2 is no longer known. The key matrices are

$$\boldsymbol{x} = \begin{bmatrix} .0025907 \\ .0028011 \\ .0033306 \\ .0036298 \\ .0041298 \end{bmatrix} \qquad F = \begin{bmatrix} 3.34711 & 0 & 0 & 0 & 0 \\ 0 & 2.60810 & 0 & 0 & 0 \\ 0 & 0 & 2.76392 & 0 & 0 \\ 0 & 0 & 0 & 2.18789 & 0 \\ 0 & 0 & 0 & 0 & 2.42639 \end{bmatrix}$$

$$K = \begin{bmatrix} 1 & 0 & 0 & 0 & 0 \\ -1 & 1 & 0 & 0 & 0 \\ 1 & -2 & 1 & 0 & 0 \\ -1 & 3 & -3 & 1 & 0 \\ 0 & -1 & 3 & -3 & 1 \end{bmatrix}.$$

This particular form for K follows the suggestion of Gersch and Kitagawa (1988). They suggested using small values for the entries in the first three rows. It turns out that in this example the posterior means are extremely sensitive to the choice of these entries. A serious graduation problem would involve some 60 to 100 rows and so the effect of the first three would be minimal. To make the numbers more appealing, the elements of F have been multiplied by 10^6. This will have the effect of multiplying the value of τ^2 by 10^6. The prior distribution $\pi_{22}(\tau^2) \propto 1$ will be used. The more reasonable prior $\pi_{22}(\tau^2) \propto 1/\tau^2$ yields an improper posterior distribution.

The posterior mean of τ^2 is found by using the density in (7.76) while the posterior mean of one the ith graduated value is found by integrating the ith element of $\boldsymbol{\mu}_1$ as given in (7.75) against the density in (7.76). All of these are one-dimensional integrations so the adaptive Gaussian procedure works quite well. The posterior mean of τ^2 is 9.2554 and the posterior means of the mortality probabilities are (.0011574, .0017584, .0024540, .0031951, .0039889). These are not particularly satisfying. The special K matrix used here produces a graduation that does not have the desirable property of having the weighted (by the reciprocal of the diagonal elements of F) average of the graduated values equal to the weighted average of the original observations. Another choice is to obtain the empirical Bayes solution, performing the graduation as in Chapter 4 but using the estimated value of τ^2. This yields (.0020616, .0026706, .0032329, .0036910, .0041122). It would also be feasible to use this value of τ^2 along with the standard K matrix as given in (4.3) to obtain graduated values that do have the desirable properties of a Whittaker graduation.

A drawback of the graduation approach as outlined here is that shrinking is forced to be toward the second level mean of $\mathbf{0}$. An alternative is to postulate a mean vector $\boldsymbol{\mu}$ based on results from earlier experience. The formulas could then be altered to reflect this change. It is not possible to estimate $\boldsymbol{\mu}$ as there is an insufficient amount of information.

9. MODIFICATIONS TO THE HIERARCHICAL NORMAL LINEAR MODEL

All the modifications discussed in this Chapter have to do with relaxing the assumption of normality. The first two cover specific distribution choices—lognormal and Poisson. In all cases the normal distribution is retained for the second level. This can usually be accomplished by careful parametrization of the first level parameters. The third modification presented is a general method for dealing with non-normal distributions.

A. LOGNORMAL

This is an especially easy case due to the simple relationship between the normal and lognormal distributions. Assume that at the first level each observation has a lognormal distribution. That is,

$$ln(X_{ij}) \sim N(\theta_i, \sigma^2/P_{ij}) \tag{9.1}$$

where X_{ij} is losses per exposure. All the customary formulas can be used once natural logarithms are taken of all of the observations. The major drawback to this model is that the mean (of X_{ij}) depends on the exposure. In particular, for predicting future observations we have to compute the following posterior expectation:

$$E(e^{\theta + \sigma^2/2P} \mid \boldsymbol{x}). \tag{9.2}$$

Again we see that the value depends on the exposure.

One way to alleviate this problem is to change the first level model to

$$ln(X_{ij}) \sim N(\theta_i - \sigma^2/2P_{ij}, \sigma^2/P_{ij}) \tag{9.3}$$

or, equivalently,

$$ln(X_{ij}) + \sigma^2/2P_{ij} \sim N(\theta_i, \sigma^2/P_{ij}). \tag{9.4}$$

The expected value of X_{ij} is now simply $exp(\theta_{ij})$. The problem now is that σ^2 must be known (or estimated) in advance.

B. POISSON

The appropriate method for handling the Poisson distribution was introduced in Chapter 6. The key is a variance stabilizing transformation. We make the additional assumption that this produces values that approximately normal. As seen in Chapter 6, the skewness is close to zero, so this is not an unreasonable statement. The use of this transformation in the *HNLM* is illustrated with Data Set 2 and an analysis that parallels the presentation in the previous Chapter.

Recall that Y_{ij} is the frequency count and it is reasonable to assume that it has the Poisson distribution with parameter $\alpha P_{ij}\lambda_i$ where αP_{ij} is the number of opportunities to have an accident (the constant factor α is an adjustment to reflect the fact that P_{ij} is only proportional to this number). λ_i is the expected number of claims per opportunity for members of the ith class. The first level observation is $x_{ij} = Y_{ij}/P_{ij}$ and the first level parameter is $\theta_i = E(x_{ij} \mid \alpha, \lambda_i) = \alpha\lambda_i$. The transformation is $z_{ij} = 2\sqrt{x_{ij}}$ and its mean and variance are approximately $\gamma_i = 2\sqrt{\theta_i}$ and $1/P_{ij}$ respectively. The one-way model can now be used with γ_i replacing θ_i and $\sigma^2 = 1$. This last replacement is especially convenient as all integrals will be one-dimensional.

Using the prior $\pi_{22}(\tau^2) \propto 1/\tau^2$ and the density in (7.48) along with adaptive Gaussian integration yields $E(\tau^2 \mid \boldsymbol{x}) = .02951$. The empirical Bayes estimate of γ_i is

$$\tilde{\gamma}_i = \frac{P_i}{P_i + 33.891}\hat{\gamma}_i + \frac{33.891}{P_i + 33.891}\hat{\mu},$$

$$\hat{\gamma}_i = \sum P_{ij} 2\sqrt{x_{ij}}/P_i, \quad \hat{\mu} = \sum \frac{P_i \hat{\gamma}_i}{1 + P_i \tau^2} \Big/ \sum \frac{P_i}{1 + P_i \tau^2}. \tag{9.5}$$

The predicted estimate of θ_i is found by taking the inverse transformation: $\tilde{\theta}_i = \tilde{\gamma}_i^2/4$. Using the same six classes as in Tables 8.8 and 8.9 the predicted frequencies for year seven are given in Table 9.1. The normal model predictions from Chapter 8 are also presented. In all six cases the Poisson model provided a superior prediction and the overall measure (8.14) is slightly better.

It is also possible to use (7.38) to obtain $Var(z_{i7} \mid \boldsymbol{x})$. With

$$Y_{i7} = P_{i7} z_{i7}^2/4$$

we have (approximately)

$$Var(Y_{i7} \mid \boldsymbol{x}) = (P_{i7}\tilde{\gamma}_i/2)^2 Var(z_{i7} \mid \boldsymbol{x}).$$

The standard deviations are also given in Table 9.1 and here is where the advantage of the Poisson model becomes clear.

Table 9.1

Predicted frequencies — Poisson and Normal Models

			Mean		*Std Dev*	
Class	P_{i7}	Y_{i7}	*Poisson*	*Normal*	*Poisson*	*Normal*
4	0.00	0	0.00	0.00	—	—
11	229.83	8	9.95	10.16	3.47	7.37
112	18,809.67	45	40.80	36.34	6.51	66.82
70	54.81	0	0.02	0.63	0.15	3.51
20	1,315.37	22	30.96	41.45	6.76	17.04
89	79.63	40	35.33	29.44	5.90	4.23
Prediction error			13.15	13.20		

C. NON-NORMAL MODELS BASED ON PARAMETER ESTIMATES

The solution has already been presented, appearing in Chapter 4 with the key equation in (4.30). As long as the distribution of the observations can be considered to a member of a parametric family, it will usually be possible to estimate the parameters by the method of maximum likelihood (the simplex method is especially good). Furthermore, asymptotic theory tells us that the estimates will have an approximate normal distribution with mean equal to the true parameter values and covariance equal to the inverse of the information matrix. Let $\boldsymbol{\theta}_i$ be the parameter for the observations from class i and let $\hat{\boldsymbol{\theta}}_i$ be its maximum likelihood estimate based on the observations from that class. The first level for the *HNLM* is then

$$\hat{\boldsymbol{\theta}}_i \sim N(\boldsymbol{\theta}_i, F_i) \tag{9.6}$$

where F_i is a known matrix estimating the covariance of the estimate.

Some caution should be attached to using this method. The quality of the solution is only as good as the quality of the distributional model being used. As well, the accuracy of any results also depends upon the degree to which the normal approximation for the distribution of maximum likelihood estimates is valid. Nevertheless, if the data are known to be non-normal, this approach ought to be better than one that assumes normality of the observations themselves.

The second level for a standard credibility analysis merely has the parameters varying around some constant mean. That is,

$$\boldsymbol{\theta}_i \sim N(\boldsymbol{\mu}, g). \tag{9.7}$$

This means that, for using the formulas of Chapter 7, the matrix A is the identity matrix. The matrix F is block diagonal with entries $F_1, \ldots, F_k$. The matrix B is a vertical stack of identity matrices each with the dimension of $\boldsymbol{\mu}$. Finally, the matrix G is block diagonal with k copies of g on the diagonal.

This makes the version I formulas particularly simple. The key quantities are

$$w_i = (F_i + g)^{-1} \tag{9.8}$$

$$\hat{\mu} = (\sum w_i)^{-1} \sum w_i \hat{\theta}_i \tag{9.9}$$

$$\theta_i^* = \hat{\theta}_i + F_i w_i (\hat{\mu} - \hat{\theta}_i) \tag{9.10}$$

$$\pi_{22}^*(g \mid \hat{\theta}) \propto (\prod \mid w_i \mid)^{1/2} \mid \sum w_i \mid^{-1/2} \times exp\left[-\sum (\hat{\theta}_i - \hat{\mu})' w_i (\hat{\theta}_i - \hat{\mu})/2\right] \pi_{22}(g). \tag{9.11}$$

This method will be illustrated with an example from Klugman (1990). Losses were collected from automobile bodily injury policies sold at five different policy limits. It is believed that these five classes have slightly different experience and so it is not appropriate to combine the losses into one sample. On the other hand, the sample sizes vary considerably and so it is possible that keeping the five samples separate will cause a considerable loss of accuracy for the classes with limited data. A credibility procedure provides the ideal solution as it effects a compromise between these two extremes. The data appear in Table 9.2 along with the relevant parameter estimates for a lognormal model. Because the data were grouped it was particular easy to obtain the parameter estimates and the estimated covariance matrix by the method of scoring (see Hogg and Klugman, 1984).

The only item that remains to be specified to use (9.8)—(9.11) is the specification of g. At the outset it appeared best to leave it totally general. However, the mode of the posterior distribution revealed that the covariance term is extremely close to zero and so all further work used a diagonal matrix. As in the analysis of the linear trend model in Chapter 8, the logarithms of the variances were used. Letting g_1 and g_2 be the diagonal elements, Gauss-Hermite integration revealed the posterior means to be .34835 and .03357 respectively. These are about ten times larger than the corresponding values found using a pseudo-estimator in Klugman (1990). The integrations also produced the posterior means of the lognormal parameters and they appear in Table 9.3.

An additional advantage over the pseudo-estimators is that variances can be computed. Also, it is likely that some feature (such as the mean) of the lognormal distribution may be desired. This can also be estimated by finding its posterior expectation rather than just inserting the parameter estimates in the expression.

Table 9.2

Losses on automobile bodily injury policies

	Limit				
Loss	*5000*	*10,000*	*25,000*	*50,000*	*100,000*
0–249	1230	415	660	950	4119
250–499	612	187	268	424	2047
500–999	546	136	311	377	2031
1000–1999	487	163	282	385	2099
2000–2999	269	93	170	220	1341
3000–3999	152	57	94	160	858
4000–4999	74				
5000–∞	491				
4000–5999		85	120	177	1047
6000–6999		21	31	38	278
7000–7999		17	46	65	324
8000–8999		13	27	31	179
9000–9999		12			
10,000–∞		77			
9000–10,999			32	68	328
11,000–11,999			7	11	97
12,000–13,999			26	17	158
14,000–15,999			18	36	168
16,000–18,999			10	25	130
19,000–20,999			11	23	106
21,000–22,999			8	7	33
23,000–24,999			6		
25,000–∞			63		
23,000–29,999				27	240
30,000–34,999				7	71
35,000–39,999				7	60
40,000–44,999				7	46
45,000–49,999				9	
50,000–∞				45	
45,000–54,999					120
55,000–59,999					35
60,000–69,999					39
70,000–79,999					51
80,000–99,999					68
100,000–∞					139
n	3861	1276	2190	3116	16,212

Table 9.2 -- Continued

$\hat{\mu}$	6.37172	6.42293	6.55376	6.55058	6.84037
$\hat{\sigma}$	1.84910	1.91864	1.91720	1.97554	1.97159
$1000Var(\hat{\mu})$	1.03736	3.36121	1.91298	1.43437	.264251
$1000Var(\hat{\sigma})$	1.03082	2.94208	1.47459	1.07189	.184696
$1000Cov(\hat{\mu},\hat{\sigma})$	−.19025	−.64931	−.33502	−.26162	−.043528

Table 9.3

Lognormal Parameter Estimates

	Limit				
	5,000	*10,000*	*25,000*	*50,000*	*100,000*
$\tilde{\mu}$	6.37153	6.41763	6.55442	6.54881	6.84107
$\tilde{\sigma}$	1.83657	1.91545	1.91456	1.98293	1.97302

APPENDIX. ALGORITHMS, PROGRAMS, AND DATA SETS

A. THE SIMPLEX METHOD OF FUNCTION MAXIMIZATION

This method (introduced by Nelder and Mead, 1965) is available for finding a local maximum of any function. No derivatives are required; all that is needed is the ability to evaluate the function. As a result it is slow but sure.

Let $\boldsymbol{x}$ be a $k \times 1$ vector and $f(\boldsymbol{x})$ be the function in question. The iterative step begins with $k+1$ points, $\boldsymbol{x}_1,\ldots,\boldsymbol{x}_{k+1}$, and the corresponding functional values, $f_1,\ldots,f_{k+1}$. Assume that the points have been ordered so that $f_1 < \cdots < f_{k+1}$. Now identify five new points. The first one, $\boldsymbol{y}_1$, is the center of $\boldsymbol{x}_2,\ldots,\boldsymbol{x}_{k+1}$. That is,

$$\boldsymbol{y}_1 = \sum_{j=2}^{k+1} \boldsymbol{x}_j / k.$$

The other four points are found as follows:

$$\begin{aligned}
\boldsymbol{y}_2 &= 2\boldsymbol{y}_1 - \boldsymbol{x}_1 \\
\boldsymbol{y}_3 &= 2\boldsymbol{y}_2 - \boldsymbol{y}_1 \\
\boldsymbol{y}_4 &= (\boldsymbol{y}_1 + \boldsymbol{y}_2)/2 \\
\boldsymbol{y}_5 &= (\boldsymbol{y}_1 + \boldsymbol{x}_1)/2.
\end{aligned} \tag{A.1}$$

Then let $g_2,\ldots,g_5$ be the corresponding functional values (the value at $\boldsymbol{y}_1$ will not be needed). Figure A.1 for the case $k = 2$ illustrates the location of these five points. The idea is to replace the worst point ($\boldsymbol{x}_1$) with one of these points. The decision process proceeds as follows (stop as soon as one of the replacements is made):

1. If $f_2 < g_2 < f_{k+1}$, replace $\boldsymbol{x}_1$ with $\boldsymbol{y}_2$.
2. If $g_2 \geq f_{k+1}$ and $g_3 > g_2$, replace $\boldsymbol{x}_1$ with $\boldsymbol{y}_3$.
3. If $g_2 \geq f_{k+1}$ and $g_3 \leq g_2$, replace $\boldsymbol{x}_1$ with $\boldsymbol{y}_2$.
4. If $f_1 < g_2 \leq f_2$ and $g_4 > g_2$, replace $\boldsymbol{x}_1$ with $\boldsymbol{y}_4$.
5. If $g_2 \leq f_1$ and $g_5 > f_1$, replace $\boldsymbol{x}_1$ with $\boldsymbol{y}_5$.
6. If no replacement has been made, replace each $\boldsymbol{x}_i$ with $(\boldsymbol{x}_i + \boldsymbol{x}_{k+1})/2$.

Figure A.1
Simplex Trial Values

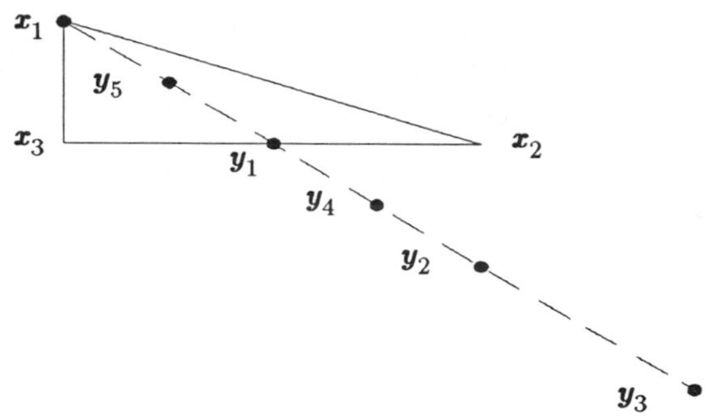

The new set of points is then re-ordered and the process repeated. In practice, do not compute $\boldsymbol{y}_3$ and g_3 until you have reached step 2. Also note that at most one of the pairs $(\boldsymbol{y}_4, g_4)$ and $(\boldsymbol{y}_5, g_5)$ needs to be obtained, depending on which (if any) of the conditions in steps 4 and 5 holds.

The only remaining items are the selection of the initial simplex and the stopping rule. One of the initial points must be obtained by intuition or a very crude grid search. An easy way to complete the initial simplex is to obtain the $i+1st$ point by increasing the ith component of $\boldsymbol{x}_1$ by a fixed percentage (say five percent). The process can be stopped when either the points or the functional values are within a certain tolerance. Nelder and Mead recommend stopping when the variance of the f's falls below a pre-set value. (While it is possible that the points on the simplex could be very dissimilar upon stopping, such an event indicates a flat likelihood surface, and in turn, unreliable estimates due to large sampling variation.)

As an example, consider the function $f(a,b) = -a^2 - 2b^2$. Table A.1 displays the first ten iterations and Figure A.2 shows the progress of the simplex.

Figure A.2
Ten Simplex Iterations

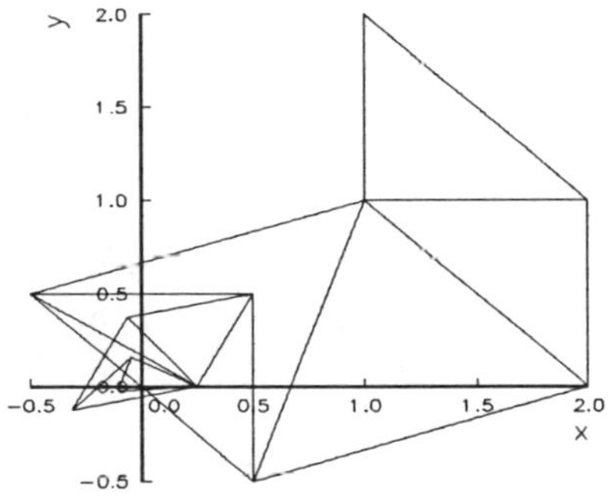

Table A.1

Ten simplex iterations

		$y_1 = (1.5,1)$	
$x_1 = (1,2)$	$f_1 = -9$	$y_2 = (2,0)$	$g_2 = -4$
$x_2 = (2,1)$	$f_2 = -6$		
$x_3 = (1,1)$	$f_3 = -3$		
		$y_1 = (1.5,.5)$	
$x_1 = (2,1)$	$f_1 = -6$	$y_2 = (1,0)$	$g_2 = -1$
$x_2 = (2,0)$	$f_2 = -4$	$y_3 = (.5,-.5)$	$g_3 = -.75$
$x_3 = (1,1)$	$f_3 = -3$		
		$y_1 = (.75,.25)$	
$x_1 = (2,0)$	$f_1 = -4$	$y_2 = (-.5,.5)$	$g_2 = -.75$
$x_2 = (1,1)$	$f_2 = -3$	$y_3 = (-1.75,.75)$	$g_3 = -4.188$
$x_3 = (.5,-.5)$	$f_3 = -.75$		
		$y_1 = (0,0)$	
$x_1 = (1,1)$	$f_1 = -3$	$y_2 = (-1,-1)$	$g_2 = -3$
$x_2 = (.5,-.5)$	$f_2 = -.75$	$y_5 = (.5,.5)$	$g_5 = -.75$
$x_3 = (-.5,.5)$	$f_3 = -.75$		
		$y_1 = (0,.5)$	
$x_1 = (.5,-.5)$	$f_1 = -.75$	$y_2 = (-.5,1.5)$	$g_2 = -4.75$
$x_2 = (-.5,.5)$	$f_2 = -.75$	$y_5 = (.25,0)$	$g_5 = -.0625$
$x_3 = (.5,.5)$	$f_3 = -.75$		
		$y_1 = (.375,.25)$	
$x_1 = (-.5,.5)$	$f_1 = -.75$	$y_2 = (1.25,0)$	$g_2 = -1.563$
$x_2 = (.5,.5)$	$f_2 = -.75$	$y_5 = (-.0625,.375)$	$g_5 = -.285$
$x_3 = (.25,0)$	$f_3 = -.0625$		
		$y_1 = (.09375,.1875)$	
$x_1 = (.5,.5)$	$f_1 = -.75$	$y_2 = (-.3125,-.125)$	$g_2 = -.129$
$x_2 = (-.0625,.375)$	$f_2 = -.285$		
$x_3 = (.25,0)$	$f_3 = -.0625$		
		$y_1 = (-.0313,-.0625)$	
$x_1 = (-.0625,.375)$	$f_1 = -.285$	$y_2 = (0,-.5)$	$g_2 = -.5$
$x_2 = (-.3125,-.125)$	$f_2 = -.129$	$y_5 = (-.0469,.1563)$	$g_5 = -.051$
$x_3 = (.25,0)$	$f_3 = -.0625$		
		$y_1 = (.1016,.0781)$	
$x_1 = (-.3125,-.125)$	$f_1 = -.129$	$y_2 = (.5156,.2813)$	$g_2 = -.424$
$x_2 = (.25,0)$	$f_2 = -.0625$	$y_5 = (-.1055,-.0234)$	$g_5 = -.012$
$x_3 = (-.0469,.1563)$	$f_3 = -.051$		
$x_1 = (.25,0)$	$f_1 = -.0625$		
$x_2 = (-.0469,.1563)$	$f_2 = -.051$		
$x_3 = (-.1055,-.0234)$	$f_3 = -.012$		

B. ADAPTIVE GAUSSIAN INTEGRATION

Ordinary Gaussian integration uses the approximation

$$\int_{-1}^{1} f(x)dx \doteq \sum_{i=0}^{n} w_i f(x_i) \tag{A.2}$$

where the weights, w_i, and the arguments, x_i are chosen so that the formula will be exact for polynomials of degree $2n+1$ or less. All the integrations done here used $n = 9$ (ten point integration). The values are given in Table A.2.

For approximating an arbitrary integral from a to b, the formula is

$$\int_{a}^{b} f(x)dx \doteq \sum_{i=0}^{n} w_i^* f(x_i^*) \tag{A.3}$$

where

$$w_i^* = (b-a)w_i/2 \quad \text{and} \quad x_i^* = (b-a)x_i/2 + (b+a)/2. \tag{A.4}$$

Table A.2

Gaussian Integration Constants

w_i	x_i
.066671344308688	−.973906528517172
.149451349050581	−.865063366688985
.219086362515982	−.679409568299024
.269266719309996	−.433395394129247
.295524334714753	−.148874338981631
.295524334714753	.148874338981631
.269266719309996	.433395394129247
.219086362515982	.679409568299024
.149451349050581	.865063366688985
.066671344308688	.973906528517172

If (A.3) does not produce a sufficiently accurate result (and measuring the accuracy is difficult when the true value is unknown) one remedy is to use a formula with more points. A more tractable solution is to split the range from a to b into smaller intervals and approximate the integral over each with the ten point formula. The results are then added to obtain the final approximation. It can be shown that by taking smaller and smaller intervals, the error can be made as small as desirable.

Using subdivisions does not solve the problem of achieving a prespecified level of accuracy. That problem is solved by the adaptive approach. Let $S(a,b)$ represent the ten point Gaussian approximation to the integral from a to b. It can be shown that (letting $m = (a+b)/2$)

$$| \int_a^b f(x)dx - S(a,m) - S(m,b) |$$

$$\doteq | S(a,b) - S(a,m) - S(m,b) | /(2^{20} - 1). \tag{A.5}$$

This means that if after splitting the interval in two the absolute difference in the two approximations is $(2^{20} - 1)\epsilon$ then the error in using the approximation based on the two intervals will be ϵ. So begin by specifying ϵ and then multiplying it by 1,000,000 (a reasonable approximation to $2^{20} - 1$). If after one division the change in approximation is less than that number, stop. Otherwise, split each of the two intervals into two smaller intervals and make the comparison for each. This time, the change for each split must be less than one-half the target used the first time in order for the total error to remain within the desired bound. Continue splitting until all subintervals meet the goal. This process is called adaptive because once the goal is satisfied for some subset of (a,b) no further subdivisions are used for that portion.

As an example, suppose we are integrating from 1 to 10 and desire the error to be less than $\epsilon = .000001$. The error goal is 1. Begin with $S(0,10) = 11.99$. Splitting the interval yields $S(0,5) = 4.48$ and $S(5,10) = 10.45$. The error is $| 11.99 - 4.48 - 10.45 | = 2.94$, which exceeds the goal. So split both of these intervals. The results for the first interval are $S(0,2.5) = 1.57$ and $S(2.5,5) = 3.81$ for an error of $| 4.48 - 1.57 - 3.81 | = .9$, which exceeds the goal of .5 (the goal must be cut in half at each new level). For the second interval, $S(5,7.5) = 4.91$ and $S(7.5,10) = 5.67$ for an error of .13, which does meet the goal. Now continue splitting the first two intervals. $S(0,1.25) = .51$ and $S(1.25,2.5) = 1.29$ for an error of .23, which meets the goal (now .25) and $S(2.5,3.75) = 1.76$ and $S(3.75,5) = 2.09$ for an error of .04 which also meets the goal. We can now stop and announce the approximate integral as $.51 + 1.29 + 1.76 + 2.09 + 4.91 + 5.67 = 16.23$.

It must be noted that these formulas deal with the absolute, and not the relative error, therefore some knowledge of the scale of the answer is necessary before this method can be used.

C. GAUSS-HERMITE INTEGRATION

The constants to use as h and w in Gauss-Hermite integration (eq. 3.10) appear in Table A.3

Table A.3

Gauss-Hermite Constants

h_i	x_i
Three-Point	
0.2954089752	−1.2247448714
1.1816359006	0
0.2954089752	1.2247448714
Four-Point	
0.0813128354	−1.6506801239
0.8049140900	−0.5246476233
0.8049140900	0.5246476233
0.0813128354	1.6506801239
Five-Point	
0.0199532421	−2.0201828705
0.3936193232	−0.9585724646
0.9453087205	0
0.3936193232	0.9585724646
0.0199532421	2.0201828705
Six-Point	
0.0045300099	−2.3506049737
0.1570673203	−1.3358490740
0.7246295952	−0.4360774119
0.7246295952	0.4360774119
0.1570673203	1.3358490740
0.0045300099	2.3506049737

D. POLAR METHOD FOR GENERATING NORMAL DEVIATES

The polar method begins with two independent uniform (0,1) random numbers, U_1 and U_2. Let $V_1 = 2U_1 - 1$ and$V_2 = 2U_2 - 1$. Next let $S = V_1^2 + V_2^2$. If $S \geq 1$, start over with two new uniform variables. Otherwise, let $T = \sqrt{-(2lnS)/S}$. Finally, let $Z_1 = V_1T$ and $Z_2 = V_2T$. Z_1 and Z_2 will be independent $N(0,1)$ random variables.

E. GAUSS PROGRAMS

The programs listed in this Section are written in the GAUSS programming language. It is a matrix language in which all variables are two-dimensional matrices. The ordinary arithmetic operations of +, −, and * all follow the usual definition for matrices. A prime (') is used to take the transpose. The operations .* and ./ mean that multiplication and division are on an element by element basis. Some other useful commands to note are SUMC(x) which sums the elements in each column of x and places the results in a column vector, ZEROS(m,n) which creates an $m \times n$ matrix of zeros, ONES($m.n$) which creates an $m \times n$ matrix of ones, EYE(m) which creates an $m \times m$ identity matrix, $x \mid y$ which concatenates the two matrices vertically, and $x\tilde{}y$ which concatenates the two matrices horizontally. All the programs are written as PROCs, another name for subroutine. The arguments are local to the procedure. Other local variables are declared in the LOCAL statement while those variables declared in the CLEARG statement are cleared as well as made global. Typically, one or more of the arguments to a procedure will be a function call. These are declared as well. Any other special GAUSS commands are noted in the annotations to the programs (Comments are enclosed between /* and */).

1. Simplex Maximization

This program maximizes a function using the simplex method. The inputs are the name of the function, a vector containing the starting values, and the tolerance for the stopping rule. The procedure will stop when the sum of the absolute relative distances between the best and second best points on the simplex is less than a specified value. The output vector contains the solution with an additional element containing the value of the function at that point.

/*SIMPLEX MAXIMIZATION
The call is
SIMPLEX(&OBJFUNC,STARTVAL,TOL)
where
&OBJFUNC is the name of the function to be maximized. The argument should be a single vector and it should return a single functional value.
STARTVAL is a column vector containing the starting value for the iterative procedure. It is essential that the dimension of this vector match the dimension of the argument to &OBJFUNC.
TOL is the tolerance for the convergence criterion. The iterations will stop when the relative distance between the two best points in the simplex is less than TOL.
The program returns a vector of length one longer than the length of STARTVAL. The last element is the value of the function at the max, the rest is the max itself.
The current best vector SIMPBEST is a global vector so if the program crashes or is aborted, it remains available.*/

```
PROC SIMPLEX(&OBJFUNC,STARTVAL,TOL);
LOCAL OBJFUNC:PROC;
CLEARG SIMPBEST;
LOCAL SIMPLEX, I, J, VV, SIMPVAL, SIMPBEST, SIMP2BST,
DECIS, MAXINDX, MAXVAL, MAX2IND, SCNDINDX, MININDX,
MINVAL, SCNDVAL, MIDPOINT, REFPOINT, REFVAL, SORTVAL,
NEWPOINT, NEWVAL, DBLPOINT, DBLVAL, HAFPOINT, HAFVAL,
RESULT;
VV=ROWS(STARTVAL);    /*No of Variables, ROWS return the number of rows of
the vector.*/
SIMPLEX=ZEROS(VV+1,VV);  /*Each row of the simplex is an argument to the
function.*/
SIMPLEX[1,.]=STARTVAL';
I=2;
DO WHILE I<=VV+1;
  J=1;
  DO WHILE J<=VV;
    SIMPLEX[I,J]=SIMPLEX[1,J];
    IF J==I-1;
      SIMPLEX[I,J]=SIMPLEX[1,J]*(1.05);
    ENDIF;    /*Each member of the initial simplex has one element of
    STARTVAL increased by 5%. Two equals signs are used when testing for
    equality. One equals sign is used for assignments.*/
    J=J+1;
  ENDO;
  I=I+1;
ENDO;
```

```
SIMPVAL=ZEROS(VV+1,1)~SEQA(1,1,VV+1); /*Functional values of each
member of the simplex are held here, the second column holds postion values for
sorting. The SEQA command creates a column vector starting at 1 with elements
increasing by 1 with VV+1 elements altogether.*/
I=1;
DO WHILE I<=VV+1;
  SIMPVAL[I,1]=OBJFUNC(SIMPLEX[I,.]'); /*Place the evaluations of
the function in a vector.*/
  I=I+1;
ENDO;
SIMPLEX~SIMPVAL[.,1]; /*Display the simplex and functional values.
Naming a variable causes it to be printed on the screen.*/
SORTVAL=SORTC(SIMPVAL,1); /*Sorts the values of the function.*/
MAXINDX=SORTVAL[VV+1,2]; /*Index of best value.*/
MAX2IND=SORTVAL[VV,2]; /*Index of second best value.*/
SIMPBEST=SIMPLEX[MAXINDX,.]'; /*The vector x sub vv+1 in the notation
of Section A.*/
SIMP2BST=SIMPLEX[MAX2IND,.]'; /*The vector x sub vv.*/
DO WHILE
  (SIMPBEST-SIMP2BST)'((SIMPBEST-SIMP2BST)
    ./SIMPBEST./SIMPBEST)>TOL*TOL; /*This is the check for
  convergence.*/
  DECIS=0; /*Needed to note that decision to end search for new value has been
  made.*/
  MININDX=SORTVAL[1,2]; /*Index of worst value.*/
  SCNDINDX=SORTVAL[2,2]; /*Index of second worst value.*/
  MIDPOINT=(SUMC(SIMPLEX)-SIMPLEX[MININDX,.]')/VV;
  /*Average of all vectors other than the worst point. This is y sub 1.*/
  REFPOINT=2*MIDPOINT-SIMPLEX[MININDX,.]'; /*Reflection of
  worst point through the midpoint. This is y sub 2.*/
  REFVAL=OBJFUNC(REFPOINT); /*This is g sub 2.*/
  IF (REFVAL<SORTVAL[VV+1,1])
    AND (REFVAL>SORTVAL[2,1]);
    DECIS=1;
    NEWPOINT=REFPOINT;
    NEWVAL=REFVAL;
  ENDIF; /*If reflected point is better than the second worst but not as good as the
  best, use it. New point and newval are the new argument and functional value for
  the next iteration.*/
  IF REFVAL>SORTVAL[VV+1,1];
    DBLPOINT=2*REFPOINT-MIDPOINT; /*If reflected point is better
    than the best point, consider a double reflection. This is y sub 3. In any event
    take the best of the two reflections as the new point.*/
    DBLVAL=OBJFUNC(DBLPOINT); /*This is g sub 3.*/
```

```
  IF DBLVAL>REFVAL;
    DECIS=3;
    NEWPOINT=DBLPOINT;
    NEWVAL=DBLVAL;
  ELSE;
    DECIS=1;
    NEWPOINT=REFPOINT;
    NEWVAL=REFVAL;
  ENDIF;
ENDIF;
IF (DECIS==0) AND (REFVAL<SORTVAL[1,1]); /*If reflected value
is worse than the worst point, consider a point halfway between the worst point and
the midpoint.*/
  HAFPOINT=(MIDPOINT+SIMPLEX[MININDX,.]')/2;  /*This is y
  sub 5.*/
ELSE; /*If the reflected value is between the worst and second worst points,
consider a point between the midpoint and the reflected point.*/
  HAFPOINT=(MIDPOINT+REFPOINT)/2;  /*This is y sub 4.*/
ENDIF;
HAFVAL=OBJFUNC(HAFPOINT);  /*This is g sub 4 or g sub 5.*/
IF  (DECIS==0)  AND  (HAFVAL>SORTVAL[1,1])  AND
(HAFVAL>REFVAL);
  DECIS=4;  /*If the new point is better than the worst point and better than
  the reflected point, use it.*/
  NEWPOINT=HAFPOINT;
  NEWVAL=HAFVAL;
  ENDIF;
IF (DECIS>0);  /*Replace worst point with the new point.*/
  SIMPLEX[MININDX,.]=NEWPOINT';
  SIMPVAL[MININDX,1]=NEWVAL;
ELSE; /*Shrink the simplex towards the best point.*/
  SIMPLEX
    =(SIMPLEX+ONES(VV+1,1)*SIMPLEX[MAXINDX,.])/2;
  I=1;
  DO WHILE I<=VV+1;
    SIMPVAL[I,1]=OBJFUNC(SIMPLEX[I,.]');
    I=I+1;
  ENDO;
ENDIF;
SIMPLEX~SIMPVAL[.,1];
SORTVAL=SORTC(SIMPVAL,1); /*Sorts the values of the function.*/
MAXINDX=SORTVAL[VV+1,2];  /*Index of best value.*/
MAX2IND=SORTVAL[VV,2];  /*Index of second best value.*/
SIMPBEST=SIMPLEX[MAXINDX,.]';
```

```
  SIMP2BST=SIMPLEX[MAX2IND,.]';      /*Two best members of the
  simplex.*/
ENDO;   /*This ends the iteration.*/
RESULT=SIMPLEX[MAXINDX,.]'|SORTVAL[VV+1,1];
RETP(RESULT);    /*This returns the vector RESULT as the output of the
procedure.*/
ENDP;
```

2. Adaptive Gaussian Integration

The program GAUSSINT performs adaptive Gaussian integration using a ten point formula. To simplify the program, the function is expected to evaluate ten arguments at once, returning ten functional values. This is done because often in GAUSS programs it is just as fast to evaluate a function at ten different points as it is to evaluate it at one point. The program also can do several integrals at once. To be suitable for Bayesian analysis, the program accepts a posterior density and then a second function which also accepts ten arguments, but produces a matrix result where each column of the matrix is some function evaluated at the ten points. If the ith column represents evaluations of g_i at ten points then the ith element of the output vector will be $\int g_i(x)f(x)dx$ where $f(x)$ is proportional to the posterior density. In order to get the constant of proportionality, one of the g functions must be the identity. The program can do integrals over infinite as well as finite ranges.

```
PROC
GAUSSINT(&POST10,&INTGRD10,LOWERLIM,UPPERLIM,SIGN,TOL1,
TOL2,MAXINT,N);
/*Performs adaptive Gaussian quadrature.
N is the number of functions to be integrated.
&POST10 is a function that evaluates the posterior density. It accepts a 10 by 1 vector
of arguments and returns a 10 by 1 vector of functional values.
&INTGRD10 is the various functions to be integrated. It accepts a 10 by vector of
arguments and returns a 10 by N vector of functional values. Each column is the
evaluation of a different function. The N functions to be integrated are the results of
POST10 times the various columns of INTGRD10.
LOWERLIM and UPPERLIM are the lower and upper limits of the integration. They
are the same for all N integrals. If the upper limit is infinity, set
UPPERLIM<LOWERLIM. LOWERLIM will still be used as the lower limit. The first
interval to be integrated over will run from LOWERLIM to 2*LOWERLIM -
UPPERLIM.
SIGN is 1 or -1. If 1, nothing is changed. If -1 the integral will be done from
-UPPERLIM to -LOWERLIM. Use this to integrate from -infinity to a constant.
```

TOL1 is the stopping rule for the adaptive step. Halving of intervals will stop when the error in the sum of the two halves versus the full interval is less than TOL1 divided by 2 to the number of halvings already done. This should yield an approximate integral that has an absolute error of about TOL1/1,000,000. When several integrals are being done, all must meet the goal in order to stop.
TOL2 is the stopping rule when the interval is infinite. New intervals will be added until the relative addition to the integral is less than TOL2. Iterations stop only when all N integrals meet the criterion.
MAXINT is the maximum number of intervals to be used.
The output is an N by 1 vector containing the integrals.*/

```
LOCAL X, W, UPPERINF, ENDPT, CONVERG, ANSWER, L, I,
  OLDINT, INTEG, ARG, ROWNUM, VECT, INSVAL, TOLVEC1,
  TOLVEC2;
LOCAL POST10:PROC, INTGRD10:PROC;
LET X = -.973906528517172 -.865063366688985
        -.679409568299024 -.433395394129247
        -.148874338981631  .148874338981631
         .433395394129247 .679409568299024
         .865063366688985 .973906528517172;
     /*vector of arguments*/
LET W = .066671344308688 .149451349050581
        .219086362515982 .269266719309996
        .295524334714753 .295524334714753
        .269266719309996 .219086362515982
        .149451349050581 .066671344308688;
     /*vector of weights*/
TOLVEC2=TOL2*ONES(N,1);
IF UPPERLIM<LOWERLIM;
  UPPERINF=1;
  UPPERLIM=2*LOWERLIM-UPPERLIM;
  ELSE;
  UPPERINF=0;
ENDIF;     /*Check for integration to infinity. If so, set the limits for the first
integration accordingly and set the flag upperinf to 1.*/
ANSWER=ZEROS(N,1);
DO UNTIL UPPERINF==2;          /*Iterate if infinite upper limit. Flag is set to
2 if it is time to stop.*/
  TOLVEC1=TOL1*ONES(1,N)/(UPPERLIM-LOWERLIM);  /*Set up for
  later halving.*/
  ENDPT=LOWERLIM|UPPERLIM;  /*Vector containing the boundaries of the
  subintervals.*/
  INTEG=ZEROS(2,N);  /*Vector containing the integrals over each of the
  subintervals.*/
  LET CONVERG=1 0;     /*Vector to indicate if an interval has met the
```

```
convergence criterion. Since there is one more boundary than intervals, the first
elements of INTEG and CONVERG are never used.*/
ARG=SIGN
  *((UPPERLIM-LOWERLIM)*X + UPPERLIM + LOWERLIM)/2;
INTEG[2,.]=W'(INTGRD10(ARG).*(POST10(ARG)*ONES(1,N)))
  *(UPPERLIM-LOWERLIM)/2;
L=2;
DO UNTIL (L>MAXINT) OR (SUMC(CONVERG)==L);
  I=2;
  DO WHILE I <= L;
    IF CONVERG[I,1]==0;   /*If interval not ok, split it in two and see if
    no improvement.*/
      OLDINT=INTEG[I,.]; /*The old value of the integral.*/
      INSVAL=0;
      ROWNUM=I-1;
      GOSUB INSERT(CONVERG); /*Lengthen the vector by inserting a
      zero in the i-1st spot. INSERT is a subroutine definied at the end of this
      program.*/
      POP CONVERG;
      GOSUB INSERT(INTEG);
      POP INTEG;
      INSVAL=(ENDPT[I-1,1]+ENDPT[I,1])/2;
      GOSUB INSERT(ENDPT); /*Insert the midpoint as a new interval
      boundary.*/
      POP ENDPT;
      L=L+1; /*There is now one more interval.*/
      ARG=SIGN*((ENDPT[I,1]-ENDPT[I-
      1,1])*X+ENDPT[I,1]+ENDPT[I-1,1])/2;
      INTEG[I,.]
        =W'(INTGRD10(ARG).*(POST10(ARG)*ONES(1,N)))
        *(ENDPT[I,1]-ENDPT[I-1,1])/2;
      ARG=SIGN*((ENDPT[I+1,1]-
      ENDPT[I,1])*X+ENDPT[I+1,1]+ENDPT[I,1])/2;
      INTEG[I+1,.]
        =W'(INTGRD10(ARG).*(POST10(ARG)*ONES(1,N)))
        *(ENDPT[I+1,1]-ENDPT[I,1])/2;   /*Integrate over the
        two new subintervals.*/
      IF ABS(INTEG[I,.]+INTEG[I+1,.]-OLDINT)
        < TOLVEC1*(ENDPT[I+1,1]-ENDPT[I-1,1]);
        CONVERG[I,1]=1;
        CONVERG[I+1,1]=1;
      ENDIF;   /*Test for two subintervals to be no improvement. If so, enter
      1's to indicate convergence, so these intervals will not be split again.*/
      I=I+1;
```

```
        ENDIF;
        I=I+1;      /*Move on to next unchecked subinterval.*/
      ENDO;
      L; /*Print L on the screen as a reminder of progress.*/
    ENDO;
    ANSWER=ANSWER+SUMC(INTEG);  /*Add up the integrals over all the
    subintervals to produce the result.*/
    UPPERLIM=3*UPPERLIM-2*LOWERLIM;
    LOWERLIM=(UPPERLIM+2*LOWERLIM)/3; /*Move limits for next interval
    when integrating to infinity.*/
    IF (UPPERINF==0)
      OR (ABS(SUMC(INTEG)./ANSWER)<TOLVEC2); /*If finite interval
      or if infinite and convergence test met, stop.*/
      UPPERINF=2;
    ENDIF;
ENDO;
RETP(ANSWER);
END;
INSERT: /*This subroutine inserts a row of INSVAL between rows ROWNUM and
ROWNUM+1 of the matrix that is the argument to INSERT when it is called. The
new matrix will have one more row and the same number of columns as the original
matrix.*/
    POP VECT;
    VECT=VECT[1:ROWNUM,.]|INSVAL*ONES(1,COLS(VECT))
      |VECT[ROWNUM+1:L,.];
    RETURN(VECT);
ENDP;
```

3. Gauss-Hermite Integration

There are three programs for performing this integration. The first is MUSIG which first uses the simplex method to find the posterior mode. It then approximates the Hessian using (3.20). The second program is GHINT1; it uses the Gauss-Hermite formula to find the posterior mean and covariance matrix. As input the number of points for the integration formula (from 3 through 6) can be selected as well as the convergence criterion. It is best to first run this with the output from MUSIG as input and with 3 point integration. The output from that run (global matrices mugh and sigmagh) can then be used as input to another run of the program, this time with 4 points. The number of points can be increased to 5 or 6 for additional accuracy if time permits. One iteration of this program requires x^y function calls where x is the number of points for the integration formula and y is the dimension of the integration. The final program is GHINT2. It simultaneously computes integrals of the form $\int g_i(\boldsymbol{x})f(\boldsymbol{x})d\boldsymbol{x}$. Unlike the Gaussian integration program, this one automatically does the integral with $g(\boldsymbol{x}) = 1$.

```
PROC (0)=MUSIG(&LNPOST,MU,STEP,TOL);     /*The (0) is a GAUSS
statement to indicate that this procedure does not return any values. This procedure
finds starting values for use with Gauss-Hermite integration and the Tierney-Kadane
method.
&LNPOST returns the logarithm of the posterior density. It accepts one argument and
returns one value.
MU is a vector of starting values for the simplex iteration to find the mode of the
posterior density.
STEP is the stepsize for the calculation of the Hessian. If 0 is selected a default value
of 0.000001 will be used. The actual value will be STEP*MUSTART.
TOL is the tolerance for the stopping rule in the simplex method. The output is two
global matrices -- MUSTART and SIGSTART that hold the results. An additional
global value is MAXP, the maximizing value from the simplex method. This allows for
rescaling when the posterior is calculated by exponentiating LNPOST.*/
CLEARG MUSTART, SIGSTART, MAXP;
LOCAL I, J, DIM, HESSMAT, H;
LOCAL LNPOST:PROC;
MUSTART = SIMPLEX(&LNPOST,MU,TOL);
DIM=ROWS(MU);   /*The dimension of MU.*/
MAXP=MUSTART[DIM+1,1];   /*The last element of MUSTART is the value of
LNPOST at the maximum.*/
MUSTART=MUSTART[1:DIM,1];     /*Selects the first DIM elements of
MUSTART.*/
HESSMAT=ZEROS(DIM,DIM);
IF STEP==0;
  STEP=1E-6;
ENDIF;
H=DIAGRV(ZEROS(DIM,DIM),STEP*MUSTART);     /*This places the
elements of STEP*MUSTART on the diagonal of a DIMxDIM matrix.*/
I=1;
DO WHILE I<=DIM;
  J=I;
  DO WHILE J<=DIM;
    HESSMAT[I,J]=(LNPOST(MUSTART+H[.,I]+H[.,J])
      -LNPOST(MUSTART+H[.,I])
      -LNPOST(MUSTART+H[.,J])+LNPOST(MUSTART))/
      H[I,I]/H[J,J];
    HESSMAT[J,I]=HESSMAT[I,J];
    J=J+1;
```

```
  ENDO;
  I=I+1;
ENDO;
SIGSTART=INV(-HESSMAT);
SIGSTART;
ENDP;
```

```
PROC (0)=GHINT1(&LNPOST,MU,SIGMA,TOL,POINTS);
```

/*This procedure produces the mean vector and covariance matrix of an arbitrary pdf. It used Gauss-Hermite integration and so the support must be the entire real line in each dimension.
&LNPOST is the logarithm of the posterior density. It accepts a single vector argument and returns one functional value. There must be a global variable present called MAXP that contains the maximizing value.
MU is a starting value for the mean. It must have the same dimension as an argument to LNPOST.
SIGMA is a starting value for the covariance matrix. It must be square, postive definite, and of the same dimension as an argument to LNPOST.
TOL is the stopping tolerance. Iterations will cease when the sum of
the relative changes in MU and SIGMA is less than TOL.
POINTS is the number of points in the Gauss-Hermite formula. It can be 3, 4, 5, or 6. The time for one iteration is proportional to the number of points raised to the dimension of the integral.
There is no direct output from this procedure. It will create two global
matrices, MUGH and SIGMAGH that contain the results.*/

```
LOCAL W, X, WW, XX, I, J, HH, D, MUH, TEST, DEN, M, MM,
    WEIGHT, ARG, FLAG, F, COUNT, S, MUS, DIM, WWW, XXX;
LOCAL LNPOST:PROC;
CLEARG MUGH, SIGMAGH;
DIM=ROWS(MU); /*The dimension of the integration.*/
LET W[4,6]=.29540897515     1.1816359006    .29540897515
           0                0               0
           .081312835447    .80491409001    .80491409001
           .081312835447    0               0
           .019953242059    .39361932315    .94530872048
           .39361932315     .019953242059   0
           .0045300099055   .15706732032    .72462959522
           .72462959522     .15706732032   .0045300099055;
LET X[4,6]= -1.2247448714    0              1.2247448714
             0               0              0
            -1.6506801239   -.5246476233    .5246476233
             1.6506801239    0              0
            -2.0201828705   -.9585724646    0
```

```
                .9585724646   2.0201828705    0
              -2.35060497367 -1.3358490740 -.4360774119
                .4360774119   1.3358490740 2.35060497367;
WW=W[POINTS-2,1:POINTS]';   /*Vector of weights.*/
XX=X[POINTS-2,1:POINTS]';   /*Vector of points.*/
HH=CHOL(SIGMA);   /*Choleski factorization of SIGMA.*/
MUH=MU;
D=DIAGRV(EYE(DIM),DIAG(HH));
HH=HH'; HH=HH*INV(D);  /*Factorization -- SIGMA = HH*D*D*HH'.*/
WW=WW.*EXP(XX.*XX);
TEST=1;
DO UNTIL TEST<TOL;
  DEN=0;
  M=ZEROS(DIM,1);
  MM=ZEROS(DIM,DIM);
  XXX=XX[1,1]*ones(DIM,1);
  WWW=WW[1,1]*ones(DIM,1);  /*The vectors to be evaluated.*/
  COUNT=ONES(DIM,1);  /*This vector keeps track of which term in each of the
  DIM summations is being evaluated.*/
  FLAG=0;
  DO WHILE FLAG==0;
    ARG=MUH+SQRT(2)*HH*D*XXX;
    F=EXP(LNPOST(ARG)-MAXP+250);
    WEIGHT=PRODC(WWW);
    DEN=DEN+F*WEIGHT;  /*Accumulating the integral with 1 as the
    integrand.*/
    M=M+ARG*F*WEIGHT;  /*Accumulating the integral for the mean.*/
    MM=MM+F*WEIGHT*ARG*ARG';  /*Accumulating the integral for the
    matrix of second moments.*/
    IF COUNT[DIM,1]<POINTS;  /*The inner summation has not yet reached
    its upper limit.*/
      COUNT[DIM,1]=COUNT[DIM,1]+1;
      WWW[DIM,1]=WW[COUNT[DIM,1],1];  /*Change the weight for the
      last factor.*/
      XXX[DIM,1]=XX[COUNT[DIM,1],1];  /*Change the last
      argument.*/
    ELSE;
      COUNT[DIM,1]=1;  /*Return to the beginning of the inner sum.*/
      WWW[DIM,1]=WW[1,1];  /*Get the first weight for the inner sum.*/
      XXX[DIM,1]=XX[1,1];  /*Change the argument.*/
      J=DIM-1;  /*Work backwards. Anytime the count is at POINTS, change
      it to 1 and replace the last weight and argument with the first weight and
      argument.*/
      DO WHILE COUNT[J,1]==POINTS;
```

```
          COUNT[J,1]=1;
          WWW[J,1]=WW[1,1];
          XXX[J,1]=XX[1,1];
          J=J-1;
          IF J==0;
            FLAG=1;
            J=1;
          ENDIF;
        ENDO;   /*When done, advance the count on the Jth dimension and change
        the weight and argument accordingly.*/
        IF FLAG==0;
          COUNT[J,1]=COUNT[J,1]+1;
          WWW[J,1]=WW[COUNT[J,1],1];
          XXX[J,1]=XX[COUNT[J,1],1];
        ENDIF;
      ENDIF;
    ENDO;
    S=MM/DEN-M*M'/(DEN*DEN);  /*Compute the covariance matrix from the
    first and second moments.*/
    HH=CHOL(S); D=DIAGRV(EYE(DIM),DIAG(HH));
    HH=HH';
    HH=HH*INV(D);
    MUS=M/DEN;
    TEST=MAXC(ABS((MUS-MUH)./MUS));
    MUH=MUS;
    MUH'; /*Print the current values to the screen.*/
ENDO;
MUGH=MUH; SIGMAGH=S';
MUGH; SIGMAGH;   /*Print the answers to the screen.*/
ENDP;


PROC (1)=GHINT2(&LNPOST,&INTGRD,POINTS,N);
/*GAUSS-HERMITE integration from final values of the mean vector and covariance
matrix. These must be in the global matrices MUGH and SIGMAGH.
&LNPOST is the logarithm of the posterior density, it returns one value from a single
argument. There must be a global variable called MAXP which holds the value of
&LNPOST at its maximum.
&INTGRD is the integrand and returns a vector allowing multiple integrals to be done
at once.
POINTS is the number of points in the Gauss-Hermite formula (3-6).
N is the number of integrals to be evaluated. The identity function need not be
specifically selected.
The output is a vector of the integrals with the posterior density properly scaled.*/
```

```
LOCAL W, X, WW, XX, WWW, XXX,  J, HH, D, MUH, DEN, M,
      DIM, FLAG, WEIGHT, ARG, F, COUNT;
LOCAL LNPOST:PROC, INTGRD:PROC;
DIM=ROWS(MUGH); /*The dimension of the integration*/
LET W[4,6]=.29540897515   1.1816359006    .29540897515
           0              0               0
           .081312835447  .80491409001    .80491409001
           .081312835447  0               0
           .019953242059  .39361932315    .94530872048
           .39361932315   .019953242059   0
           .0045300099055 .15706732032    .72462959522
           .72462959522   .15706732032   .0045300099055;
LET X[4,6]= -1.2247448714   0             1.2247448714
             0              0             0
            -1.6506801239  -.5246476233   .5246476233
             1.6506801239   0             0
            -2.0201828705  -.9585724646   0
              .9585724646  2.0201828705   0
            -2.35060497367 -1.3358490740 -.4360774119
              .4360774119  1.3358490740  2.35060497367;

WW=W[POINTS-2,1:POINTS]';     /*Vector of weights.*/
XX=X[POINTS-2,1:POINTS]';     /*Vector of points.*/
HH=CHOL(SIGMAGH);
MUH=MUGH;
D=DIAGRV(EYE(DIM),DIAG(HH));
HH=HH';
HH=HH*INV(D); /*Factorization -- SIGMA = HH*D*D*HH'.*/
WW=WW.*EXP(XX.*XX);
DEN=0
M=ZEROS(N,1);
XXX=XX[1,1]*ONES(DIM,1);
WWW=WW[1,1]*ONES(DIM,1); /*The vectors to be evaluated.*/
COUNT=ONES(DIM,1); /*This vector keeps track of which term in each of the
DIM summations is being evaluated. */
FLAG=0;
DO WHILE FLAG==0;
  ARG=MUH+SQRT(2)*HH*D*XXX;
  F=EXP(POST(ARG)-MAXP+250);
  WEIGHT=PRODC(WWW);
  DEN=DEN+F*WEIGHT; /*Accumulating the integral with 1 as the integrand.*/
  M=M+INTGRD(ARG)*F*WEIGHT; /*Accumulating the integrals.*/
  IF COUNT[DIM,1]<POINTS; /*The inner summation has not yet reached its
  upper limit.*/
```

```
    COUNT[DIM,1]=COUNT[DIM,1]+1;
    WWW[DIM,1]=WW[COUNT[DIM,1],1]; /*Change the weight for the last
    factor.*/
    XXX[DIM,1]=XX[COUNT[DIM,1],1]; /*Change the last argument.*/
  ELSE;
    COUNT[DIM,1]=1; /*Return to the beginning of the inner sum.*/
    WWW[DIM,1]=WW[1,1]; /*Get the first weight for the inner sum.*/
    XXX[DIM,1]=XX[1,1]; /*Change the argument.*/
    J=DIM-1;   /*Work backwards. Anytime the count is at POINTS, change it
    to 1 and replace the last weight and argument with the first weight and
    argument.*/
    DO WHILE COUNT[J,1]==POINTS;
      COUNT[J,1]=1;
      WWW[J,1]=WW[1,1];
      XXX[J,1]=XX[1,1];
      J=J-1;
      IF J==0;
        FLAG=1;
        J=1;
      ENDIF;
    ENDO;   /*When done, advance the count on the Jth dimension and change the
    weight and argument accordingly.*/
    IF FLAG==0;
      COUNT[J,1]=COUNT[J,1]+1;
      WWW[J,1]=WW[COUNT[J,1],1];
      XXX[J,1]=XX[COUNT[J,1],1];
    ENDIF;
  ENDIF;
ENDO;
MUH=M/DEN;
RETP(MUH);
ENDP;
```

4. Monte Carlo Integration

There are two programs here. One is for aiding in determining how many simulations need to be done to achieve a desired level of accuracy. The second one performs the integration. GAUSS has a built-in normal random generator, so no extra work is required. The exponentiation of LNPOST is done inside these programs as well as the adjustment for overflow.

```
PROC MCERROR(&LNPOST,MU,TOL,DF,NUMSIM,K);
/*Determines the sample size to have a 50% chance of achieving K significant digits
when doing monte carlo integration.
&LNPOST is the logarithm of the posterior density.
MU is a starting value for the posterior mode.
TOL is the tolerance for the simplex method.
DF is the degrees of freedom for the t-distribution.
NUMSIM is the number of simulations.
The output is the required sample size.*/
LOCAL H, ANSWER, NUM, DEN, I, INVSIG, U, W, XI, T;
LOCAL LNPOST:PROC;
DIM=ROWS(MU);
CALL MUSIG(&LNPOST,MU,0,TOL);    /*Creates global variables MUSTART,
the posterior mean and SIGSTART, the inverse negative Hessian.  The program
SIMPLEX is called by MUSIG, so it must be available.*/
INVSIG=INVPD(SIGSTART);    /*INVPD finds the inverse of a positive definite
matrix.*/
H=CHOL(SIGSTART);    /*Choleski factorization.*/
H=H';
NUM=0;
DEN=0;
I=1;
DO WHILE I<=NUMSIM;
  U=RNDN(DIM,1);    /*A DIMx1 vector of independent unit normals.*/
  W=RNDN(DF,1);    /*A DFx1 vector of independent unit normals.*/
  XI=H*U*SQRT(DF/(W'W));
  T=-(DIM+DF)*LN(1+XI'INVSIG*XI/DF)/2;
  XI=XI+MUSTART;    /*A random multiviate t.*/
  T=LNPOST(XI)-T-MAXP+250;    /*This scales T to be less than 250 so there
  will no overflow.*/
  IF T>-250;    /*This avoids underflow, taking EXP(T)=0 when T is small.*/
    T=EXP(T);
    NUM=NUM+T*T;
    DEN=DEN+T;
  ENDIF;
  I=I+1;
ENDO;
ANSWER=NUMSIM*NUM/(DEN*DEN)-1;
ANSWER=ANSWER*1.8225*(10^(2*K));
RETP(ANSWER);
ENDP;
```

```
PROC MCINT(&LNPOST,&INTGRD,MU,TOL,DF,NUMSIM,N);
/*Performs Monte Carlo integration.
&LNPOST is the logarithm of the posterior density.
&INTGRD is the function to be integrated. It accepts one argument and returns a
vector with all of the integrands to be multiplied by the posterior. The constant
function need not be specifically selected.
MU is a starting value for the posterior mode.
TOL is the tolerance for the simplex method.
DF is the degrees of freedom for the t-distribution.
NUMSIM is the number of simulations.
N is the number of intgegrals.
The output is a vector of integrals.*/
LOCAL H, ANSWER, DEN, I, INVSIG, U, W, XI, T, YY;
LOCAL LNPOST:PROC, INTGRD:PROC;
DIM=ROWS(MU);
CALL MUSIG(&LNPOST,MU,0,TOL);
INVSIG=INVPD(SIGSTART);
H=CHOL(SIGSTART);
H=H';
ANSWER=ZEROS(N,1);
DEN=0;
I=1;
DO WHILE I<=NUMSIM;
  U=RNDN(DIM,1);
  W=RNDN(DF,1);
  XI=H*U*SQRT(DF/(W'W));
  T=-(DIM+DF)*LN(1+XI'INVSIG*XI/DF)/2;
  XI=XI+MUSTART;
  T=LNPOST(XI)-T-MAXP+500;    /*The adjustment of subtracting MAXP
  and adding 500 guarantees that T will be less than 500, preventing an overflow.*/
  IF T>-500;   /*If T<-500 there will be an underflow, so EXP(T) is just taken
  to be zero in this case.*/
    T=EXP(T);
    ANSWER=ANSWER+INTGRD(XI)*T;
    DEN=DEN+T;
  ENDIF;
  I=I+1;
ENDO;
ANSWER=ANSWER/DEN;
RETP(ANSWER);
ENDP;
```

5. Tierney-Kadane Integration

The following program finds the posterior mean and covariance.

```
PROC TKINT(&POSTINT,MU,TOL,STEP,N);
/*This procedure finds the Tierney-Kadane approximate integrals.
&POSTINT returns the logarithm of the posterior density + the logarithm of the
integrand. Several integrands may be done at once but the first one must be the log of
one; that is, just the log of the posterior density.
MU is a vector of starting values for the simplex iteration to find the mode of the
posterior density.
STEP is the stepsize for the calculation of the Hessian. If 0 is selected a default value
of 0.000001 will be used. The actual value will be STEP*MUSTART.
TOL is the tolerance for the stopping rule in the simplex method.
N is the number of integrals to be evaluated.
The output is a vector containing the approximate integrals.*/
LOCAL MAXP, MUSTART, SIGSTART, P1, P2, P3, P4, K, DEN,
      NUM, I, J, DIM, MUK, HESSMAT, H;
LOCAL POSTINT:PROC;
MUSTART = SIMPLEXI(&POSTINT,MU,TOL,1);     /*A special version of
the Simplex method that accepts vector-valued functions. The fourth entry of the
procedure call indicates which argument of the output vector is to be maximized. The
only modification to the program SIMPLEX that needs to be made is that each time
the fuction is called, the Kth element of the output vector must be extracted. Here the
first element is called for as at this step the maximum is being done with respect to the
log of the posterior.*/
DIM=ROWS(MU);
MAXP=MUSTART[DIM+1,1];
MUSTART=MUSTART[1:DIM,1];
HESSMAT=ZEROS(DIM,DIM);
IF STEP==0;
  STEP=1E-6;
ENDIF;
H=DIAGRV(ZEROS(DIM,DIM),STEP*MUSTART);
I=1;
DO WHILE I<=DIM;
  J=I;
  DO WHILE J<=DIM;
    P1=POSTINT(MUSTART+H[.,I]+H[.,J]);
    P2=POSTINT(MUSTART+H[.,I]);
    P3=POSTINT(MUSTART+H[.,J]);
    P4=POSTINT(MUSTART);
    HESSMAT[I,J]=(P1[1,1]-P2[1,1]-P3[1,1]+P4[1,1])
      /H[I,I]/H[J,J];
```

```
    HESSMAT[J,I]=HESSMAT[I,J];
    J=J+1;
  ENDO;
  I=I+1;
ENDO;
SIGSTART=INV(-HESSMAT);
SIGSTART;
DEN=SQRT(DET(SIGSTART));
NUM=ZEROS(N-1,1);
K=2;
DO WHILE K<=N;
  MUK=SIMPLEXI(&POSTINT,MUSTART,TOL,K);
  NUM[K-1,1]=EXP(MUK[DIM+1,1]-MAXP);
  MUK=MUK[1:DIM,1];
  H=DIAGRV(ZEROS(DIM,DIM),STEP*MUK);
  I=1;
  DO WHILE I<=DIM;
    J=I;
    DO WHILE J<=DIM;
      P1=POSTINT(MUK+H[.,I]+H[.,J]);
      P2=POSTINT(MUK+H[.,I]);
      P3=POSTINT(MUK+H[.,J]);
      P4=POSTINT(MUK);
      HESSMAT[I,J]=(P1[K,1]-P2[K,1]-P3[K,1]+P4[K,1])
        /H[I,I]/H[J,J];
      HESSMAT[J,I]=HESSMAT[I,J];
      J=J+1;
    ENDO;
    I=I+1;
  ENDO;
  SIGSTART=INV(-HESSMAT);
  NUM[K-1,1]=NUM[K-1,1]*SQRT(DET(SIGSTART));
  K=K+1;
ENDO;
NUM=NUM/DEN;
RETP(NUM);
ENDP;
```

F. DATA SETS

1. Data Set 1

i	X_{i1}	X_{i2}	X_{i3}	X_{i4}	X_{i5}	θ_i	$\overline{X}_i$
1	124.93	110.67	106.93	104.05	101.60	111.97	109.636
2	97.16	89.28	102.88	111.10	105.78	98.87	101.240
3	103.57	86.82	92.49	87.99	83.80	90.98	90.934
4	119.53	125.92	98.05	117.57	94.17	108.44	111.048
5	95.68	110.43	83.59	110.54	101.51	99.06	100.350
6	102.04	93.60	106.12	98.70	108.55	95.44	101.802
7	112.71	101.64	106.50	111.71	100.59	105.09	106.630
8	119.16	111.54	127.24	115.02	115.06	111.99	117.604
9	81.71	90.45	91.51	84.74	88.95	95.99	87.472
10	102.51	123.81	113.83	94.97	90.51	105.60	105.126

2. Data Sets 2—4

These data sets appear on the following pages. The values for Data Set 3 are the original numbers. Those in Table 8.3 had the payroll and losses divided by 10,000,000.

Data Set 2

Class	Year	Payroll	Loss	Class	Year	Payroll	Loss
1	1	32.322	1	7	1	0.000	0
1	2	33.779	4	7	2	0.000	0
1	3	43.548	3	7	3	0.000	0
1	4	46.686	5	7	4	0.000	0
1	5	34.713	1	7	5	0.000	0
1	6	32.857	3	7	6	0.000	0
1	7	36.600	4	7	7	0.000	0
2	1	45.995	3	8	1	41.403	3
2	2	37.888	1	8	2	34.066	11
2	3	34.581	0	8	3	32.729	4
2	4	28.298	0	8	4	32.235	0
2	5	45.265	2	8	5	32.777	2
2	6	39.945	0	8	6	28.620	4
2	7	39.322	4	8	7	24.263	1
3	1	289.047	5	9	1	6.452	0
3	2	392.176	8	9	2	6.927	1
3	3	368.982	8	9	3	5.851	0
3	4	323.770	8	9	4	3.033	1
3	5	385.222	16	9	5	1.787	0
3	6	346.390	8	9	6	1.074	1
3	7	324.132	9	9	7	0.685	0
4	1	0.000	0	10	1	537.311	11
4	2	0.000	0	10	2	569.041	12
4	3	0.000	0	10	3	597.146	15
4	4	0.037	0	10	4	570.295	15
4	5	0.000	0	10	5	1116.750	18
4	6	0.000	0	10	6	774.454	10
4	7	0.000	0	10	7	534.953	11
5	1	310.389	15	11	1	149.683	6
5	2	292.464	10	11	2	157.947	6
5	3	262.560	18	11	3	174.549	5
5	4	273.257	25	11	4	181.317	10
5	5	372.730	27	11	5	202.066	13
5	6	263.443	20	11	6	187.564	7
5	7	355.826	28	11	7	229.830	8
6	1	291.784	27	12	1	609.467	20
6	2	267.439	44	12	2	645.375	20
6	3	288.555	40	12	3	667.384	27
6	4	308.364	31	12	4	573.144	15
6	5	387.095	40	12	5	782.643	38
6	6	351.445	34	12	6	478.749	8
6	7	342.518	37	12	7	454.967	13

Data Set 2

Class	Year	Payroll	Loss	Class	Year	Payroll	Loss
13	1	120.027	0	19	1	77.909	3
13	2	131.020	1	19	2	78.925	0
13	3	161.145	8	19	3	77.558	1
13	4	182.135	5	19	4	69.525	1
13	5	276.520	11	19	5	65.972	0
13	6	158.310	2	19	6	63.274	0
13	7	168.420	7	19	7	75.335	2
14	1	196.722	13	20	1	1787.463	50
14	2	209.923	12	20	2	2027.230	90
14	3	196.199	13	20	3	1853.812	54
14	4	188.820	14	20	4	1742.135	63
14	5	202.807	19	20	5	2119.499	56
14	6	180.979	8	20	6	1545.169	35
14	7	195.628	6	20	7	1315.368	22
15	1	215.898	11	21	1	258.885	10
15	2	224.229	11	21	2	314.825	10
15	3	224.306	10	21	3	299.999	10
15	4	212.232	6	21	4	314.081	11
15	5	245.198	11	21	5	258.487	17
15	6	203.698	16	21	6	235.082	8
15	7	210.496	15	21	7	230.142	8
16	1	57.265	4	22	1	86.381	4
16	2	62.662	3	22	2	95.036	4
16	3	66.984	6	22	3	134.285	4
16	4	63.211	3	22	4	113.491	8
16	5	77.680	4	22	5	141.717	0
16	6	61.944	3	22	6	128.956	0
16	7	44.195	3	22	7	105.489	1
17	1	347.835	22	23	1	766.953	7
17	2	326.396	7	23	2	912.516	8
17	3	307.978	13	23	3	866.413	9
17	4	350.914	20	23	4	898.687	15
17	5	546.047	23	23	5	1806.752	34
17	6	410.980	8	23	6	1018.684	14
17	7	377.287	18	23	7	984.843	11
18	1	0.000	0	24	1	62.153	1
18	2	0.000	0	24	2	84.116	1
18	3	0.000	0	24	3	75.524	1
18	4	0.000	0	24	4	109.237	1
18	5	0.000	0	24	5	119.034	2
18	6	0.000	0	24	6	116.794	1
18	7	0.000	0	24	7	136.571	3

Data Set 2

Class	Year	Payroll	Loss	Class	Year	Payroll	Loss
25	1	3431.494	59	31	1	133.124	6
25	2	3882.069	68	31	2	156.835	13
25	3	3805.563	76	31	3	256.429	20
25	4	3919.527	77	31	4	253.184	9
25	5	4352.809	94	31	5	157.622	12
25	6	3949.550	73	31	6	75.804	2
25	7	3927.784	73	31	7	71.884	4
26	1	32.582	2	32	1	220.004	13
26	2	24.922	2	32	2	206.989	15
26	3	30.317	2	32	3	222.724	15
26	4	29.732	2	32	4	268.898	12
26	5	49.201	0	32	5	291.268	39
26	6	34.379	4	32	6	290.797	22
26	7	21.536	0	32	7	244.777	20
27	1	131.306	1	33	1	3.769	0
27	2	130.015	2	33	2	1.985	0
27	3	138.982	1	33	3	3.108	0
27	4	131.359	5	33	4	3.046	0
27	5	187.921	6	33	5	14.316	0
27	6	148.775	3	33	6	10.908	1
27	7	134.179	3	33	7	10.516	0
28	1	103.473	3	34	1	283.926	8
28	2	102.280	0	34	2	217.928	15
28	3	111.964	1	34	3	229.275	18
28	4	115.853	2	34	4	224.755	0
28	5	164.759	3	34	5	427.331	4
28	6	138.420	3	34	6	384.840	1
28	7	123.096	0	34	7	420.079	1
29	1	129.504	5	35	1	9.159	0
29	2	150.853	13	35	2	4.492	0
29	3	180.207	8	35	3	7.229	0
29	4	226.038	21	35	4	9.169	0
29	5	499.272	33	35	5	8.106	0
29	6	344.642	13	35	6	12.594	0
29	7	436.146	21	35	7	8.913	0
30	1	18.708	0	36	1	981.952	22
30	2	16.761	2	36	2	527.044	17
30	3	9.341	0	36	3	533.930	21
30	4	11.269	0	36	4	451.839	19
30	5	2.485	0	36	5	352.529	20
30	6	2.056	0	36	6	230.163	16
30	7	1.556	0	36	7	243.936	19

Data Set 2

Class	Year	Payroll	Loss	Class	Year	Payroll	Loss
37	1	543.295	12	43	1	1348.008	48
37	2	540.957	20	43	2	1408.717	59
37	3	515.061	22	43	3	1616.470	65
37	4	534.942	12	43	4	1485.584	73
37	5	620.379	13	43	5	1954.211	91
37	6	553.122	17	43	6	1528.900	62
37	7	510.214	9	43	7	1252.694	43
38	1	1029.563	67	44	1	1123.507	43
38	2	1097.433	58	44	2	1148.065	37
38	3	1272.550	68	44	3	1198.427	36
38	4	1357.884	92	44	4	1258.329	56
38	5	1622.744	68	44	5	1290.727	70
38	6	1160.147	77	44	6	1213.667	26
38	7	1056.315	57	44	7	1209.309	47
39	1	239.966	26	45	1	1558.521	50
39	2	242.634	27	45	2	1798.188	54
39	3	295.803	31	45	3	1804.204	48
39	4	281.269	25	45	4	1963.514	81
39	5	376.989	36	45	5	2758.736	122
39	6	234.967	20	45	6	2266.418	55
39	7	218.695	18	45	7	2135.638	77
40	1	29.923	1	46	1	1677.159	84
40	2	33.448	3	46	2	1818.022	132
40	3	32.704	2	46	3	1716.882	134
40	4	40.210	2	46	4	1828.931	94
40	5	53.806	0	46	5	2389.001	141
40	6	45.562	1	46	6	1838.996	93
40	7	45.396	0	46	7	1586.483	106
41	1	1100.645	54	47	1	107.175	0
41	2	1241.328	54	47	2	95.723	2
41	3	1185.711	48	47	3	78.302	0
41	4	1251.523	35	47	4	37.141	0
41	5	1420.391	44	47	5	33.023	1
41	6	1281.866	82	47	6	31.017	0
41	7	1170.673	30	47	7	43.081	0
42	1	43.101	1	48	1	113.138	5
42	2	34.826	0	48	2	148.388	15
42	3	47.346	2	48	3	199.211	9
42	4	70.784	2	48	4	245.298	12
42	5	26.081	0	48	5	305.085	13
42	6	77.421	0	48	6	282.832	14
42	7	7.120	2	48	7	179.536	15

Data Set 2

Class	Year	Payroll	Loss	Class	Year	Payroll	Loss
49	1	389.615	11	55	1	161.352	19
49	2	352.236	7	55	2	232.134	16
49	3	465.380	12	55	3	145.447	14
49	4	458.031	12	55	4	178.502	12
49	5	686.360	12	55	5	220.084	9
49	6	707.343	9	55	6	111.862	2
49	7	499.189	16	55	7	119.294	10
50	1	85.845	2	56	1	15.655	0
50	2	85.783	0	56	2	18.035	0
50	3	104.434	0	56	3	23.572	1
50	4	99.875	2	56	4	26.767	0
50	5	103.061	0	56	5	0.201	0
50	6	96.433	2	56	6	21.955	0
50	7	78.010	3	56	7	23.234	3
51	1	14.229	0	57	1	117.163	6
51	2	13.704	0	57	2	134.122	13
51	3	15.445	1	57	3	139.911	6
51	4	25.915	0	57	4	160.603	8
51	5	38.275	0	57	5	188.862	10
51	6	27.823	0	57	6	210.026	10
51	7	18.472	0	57	7	204.768	3
52	1	89.758	4	58	1	150.296	0
52	2	118.457	1	58	2	166.081	3
52	3	126.104	3	58	3	155.572	0
52	4	106.947	5	58	4	150.686	0
52	5	117.231	7	58	5	225.843	3
52	6	101.401	2	58	6	73.296	1
52	7	57.373	6	58	7	191.223	1
53	1	149.060	1	59	1	273.622	2
53	2	189.778	2	59	2	312.386	4
53	3	152.892	3	59	3	384.966	7
53	4	186.371	3	59	4	408.276	4
53	5	245.861	3	59	5	748.580	11
53	6	216.972	3	59	6	426.061	5
53	7	104.665	6	59	7	376.149	3
54	1	0.000	0	60	1	196.698	4
54	2	0.000	0	60	2	217.671	3
54	3	0.000	0	60	3	239.066	6
54	4	0.075	0	60	4	235.468	5
54	5	0.000	0	60	5	366.125	3
54	6	0.000	0	60	6	280.203	4
54	7	0.000	0	60	7	202.221	2

Data Set 2

Class	Year	Payroll	Loss	Class	Year	Payroll	Loss
61	1	0.000	0	67	1	113.122	2
61	2	0.000	0	67	2	141.183	6
61	3	0.288	0	67	3	186.196	12
61	4	0.433	1	67	4	214.230	12
61	5	1.312	0	67	5	193.758	10
61	6	1.268	0	67	6	221.813	10
61	7	0.806	0	67	7	201.130	8
62	1	128.762	6	68	1	7.329	0
62	2	129.566	5	68	2	13.180	0
62	3	156.728	2	68	3	8.650	0
62	4	170.342	4	68	4	7.822	1
62	5	205.975	5	68	5	13.752	0
62	6	161.752	1	68	6	9.581	0
62	7	128.477	1	68	7	10.006	1
63	1	3.119	0	69	1	174.102	2
63	2	3.685	0	69	2	201.453	3
63	3	3.764	0	69	3	209.368	5
63	4	3.831	0	69	4	253.491	6
63	5	4.993	1	69	5	265.070	2
63	6	3.780	0	69	6	269.752	5
63	7	2.618	0	69	7	192.365	8
64	1	28.386	0	70	1	38.714	0
64	2	32.964	1	70	2	40.550	0
64	3	44.084	0	70	3	48.034	0
64	4	81.340	2	70	4	47.987	0
64	5	89.199	1	70	5	58.351	0
64	6	81.750	0	70	6	54.275	0
64	7	87.903	2	70	7	54.812	0
65	1	91.605	1	71	1	83.008	3
65	2	112.329	1	71	2	82.796	3
65	3	90.277	2	71	3	99.997	0
65	4	115.104	3	71	4	106.843	5
65	5	106.548	2	71	5	145.506	4
65	6	89.530	3	71	6	110.815	5
65	7	101.780	1	71	7	103.664	5
66	1	64.214	0	72	1	121.196	10
66	2	81.465	0	72	2	148.555	6
66	3	66.600	0	72	3	186.487	6
66	4	82.745	0	72	4	309.997	21
66	5	94.381	1	72	5	265.607	14
66	6	86.545	1	72	6	228.599	10
66	7	92.120	1	72	7	107.859	7

Data Set 2

Class	Year	Payroll	Loss	Class	Year	Payroll	Loss
73	1	66.966	2	79	1	263.738	28
73	2	65.204	5	79	2	308.347	27
73	3	79.769	2	79	3	296.709	18
73	4	73.765	0	79	4	303.772	23
73	5	108.141	6	79	5	377.930	39
73	6	84.860	3	79	6	185.310	14
73	7	90.598	4	79	7	162.716	26
74	1	820.970	64	80	1	1013.867	39
74	2	1071.589	57	80	2	956.509	35
74	3	851.567	65	80	3	1041.898	39
74	4	772.685	41	80	4	1122.146	34
74	5	1126.973	65	80	5	1400.163	51
74	6	416.385	36	80	6	1034.801	41
74	7	308.005	16	80	7	891.256	56
75	1	259.342	10	81	1	111.090	9
75	2	282.100	31	81	2	106.213	10
75	3	265.376	34	81	3	102.541	7
75	4	254.058	27	81	4	113.139	12
75	5	335.424	33	81	5	168.177	13
75	6	244.496	23	81	6	149.237	10
75	7	214.138	20	81	7	89.148	13
76	1	646.899	31	82	1	2913.910	82
76	2	706.321	27	82	2	2941.158	106
76	3	699.304	25	82	3	3413.532	81
76	4	705.517	34	82	4	3402.478	104
76	5	867.373	46	82	5	4313.542	132
76	6	638.322	38	82	6	3024.050	96
76	7	572.799	56	82	7	2844.038	89
77	1	107.353	5	83	1	3175.937	224
77	2	104.486	8	83	2	3058.315	148
77	3	124.194	12	83	3	3226.647	160
77	4	175.424	7	83	4	3575.413	199
77	5	206.059	12	83	5	4517.677	228
77	6	177.489	7	83	6	3478.879	178
77	7	167.093	9	83	7	2593.355	178
78	1	77.171	3	84	1	79.582	4
78	2	76.506	6	84	2	58.306	7
78	3	98.673	6	84	3	100.149	11
78	4	103.028	13	84	4	78.744	10
78	5	122.334	16	84	5	231.495	24
78	6	57.534	4	84	6	276.840	11
78	7	80.913	12	84	7	459.901	34

Data Set 2

Class	Year	Payroll	Loss	Class	Year	Payroll	Loss
85	1	84.886	9	91	1	1328.932	31
85	2	94.043	3	91	2	1334.186	33
85	3	98.463	14	91	3	1334.642	32
85	4	132.101	13	91	4	1439.621	55
85	5	211.462	17	91	5	1900.358	57
85	6	172.669	28	91	6	1575.137	33
85	7	194.281	19	91	7	1495.906	39
86	1	0.000	0	92	1	1283.025	54
86	2	21.319	0	92	2	1337.012	61
86	3	20.923	0	92	3	1355.122	68
86	4	18.530	0	92	4	1519.127	91
86	5	27.521	1	92	5	1963.594	53
86	6	10.468	0	92	6	1524.377	80
86	7	11.961	0	92	7	1419.202	76
87	1	28.988	3	93	1	306.705	3
87	2	21.153	8	93	2	314.853	1
87	3	19.404	4	93	3	357.296	0
87	4	13.989	1	93	4	439.514	4
87	5	5.164	1	93	5	602.841	4
87	6	10.733	0	93	6	554.907	3
87	7	3.488	0	93	7	571.591	4
88	1	7.247	0	94	1	209.037	1
88	2	8.407	0	94	2	249.768	3
88	3	10.825	0	94	3	242.952	3
88	4	6.140	0	94	4	223.889	3
88	5	18.044	1	94	5	286.412	12
88	6	6.229	1	94	6	247.946	6
88	7	3.670	0	94	7	246.528	4
89	1	106.135	30	95	1	598.297	12
89	2	93.503	36	95	2	633.966	15
89	3	96.455	26	95	3	650.776	12
89	4	118.534	73	95	4	761.772	12
89	5	112.757	54	95	5	863.569	27
89	6	93.584	48	95	6	784.155	15
89	7	79.629	40	95	7	748.876	18
90	1	76.405	6	96	1	511.401	41
90	2	80.802	1	96	2	609.635	27
90	3	82.884	9	96	3	518.000	11
90	4	101.123	11	96	4	561.834	9
90	5	164.261	6	96	5	617.042	10
90	6	95.666	7	96	6	448.895	6
90	7	89.050	4	96	7	501.484	8

Data Set 2

Class	Year	Payroll	Loss	Class	Year	Payroll	Loss
97	1	591.968	26	103	1	183.162	12
97	2	609.840	23	103	2	204.070	11
97	3	641.186	32	103	3	215.898	15
97	4	673.982	41	103	4	254.076	22
97	5	801.838	40	103	5	309.034	16
97	6	592.050	44	103	6	270.101	21
97	7	516.338	24	103	7	244.806	24
98	1	3307.708	46	104	1	49.076	6
98	2	3156.211	42	104	2	46.452	3
98	3	3223.638	65	104	3	53.448	3
98	4	3269.966	60	104	4	45.917	0
98	5	4895.145	103	104	5	66.054	0
98	6	3694.214	75	104	6	52.828	3
98	7	3564.236	55	104	7	43.959	1
99	1	1055.156	25	105	1	2126.661	63
99	2	1065.292	30	105	2	2273.041	63
99	3	1141.624	35	105	3	2325.099	77
99	4	1193.582	48	105	4	2452.336	61
99	5	1615.875	61	105	5	3209.569	93
99	6	1337.935	49	105	6	2439.080	68
99	7	1269.005	38	105	7	2236.104	62
100	1	309.501	10	106	1	39.867	1
100	2	347.849	16	106	2	39.843	2
100	3	339.384	16	106	3	42.824	1
100	4	323.840	11	106	4	37.455	1
100	5	410.613	15	106	5	54.205	3
100	6	325.568	8	106	6	40.643	0
100	7	301.404	9	106	7	37.574	2
101	1	1768.157	12	107	1	231.292	0
101	2	1809.730	18	107	2	250.281	1
101	3	2096.204	20	107	3	241.722	1
101	4	2053.998	20	107	4	245.193	0
101	5	2942.043	25	107	5	448.444	0
101	6	2383.743	30	107	6	341.276	0
101	7	2327.415	27	107	7	295.566	0
102	1	317.952	6	108	1	679.884	14
102	2	318.712	18	108	2	906.121	17
102	3	305.967	16	108	3	950.964	7
102	4	342.672	7	108	4	977.763	9
102	5	463.343	19	108	5	1340.554	20
102	6	340.908	13	108	6	1026.951	24
102	7	329.476	10	108	7	969.722	31

Data Set 2

Class	Year	Payroll	Loss	Class	Year	Payroll	Loss
109	1	53.708	3	115	1	142.476	2
109	2	67.571	2	115	2	153.847	0
109	3	63.940	5	115	3	172.002	4
109	4	71.650	1	115	4	186.035	5
109	5	98.161	4	115	5	202.441	1
109	6	79.249	7	115	6	185.968	2
109	7	59.117	3	115	7	191.504	4
110	1	44.035	1	116	1	818.162	6
110	2	48.523	2	116	2	831.213	1
110	3	57.823	3	116	3	911.948	18
110	4	45.613	3	116	4	1031.709	13
110	5	72.047	1	116	5	1439.935	21
110	6	30.096	1	116	6	1184.156	8
110	7	24.968	2	116	7	1254.184	9
111	1	63.785	6	117	1	471.076	2
111	2	67.671	3	117	2	261.442	6
111	3	77.918	2	117	3	439.570	10
111	4	82.328	6	117	4	421.738	2
111	5	105.884	4	117	5	842.236	6
111	6	75.422	5	117	6	96.233	4
111	7	70.851	10	117	7	72.651	0
112	1	12075.618	29	118	1	90.679	2
112	2	12618.147	24	118	2	89.372	3
112	3	13622.255	21	118	3	75.521	3
112	4	14833.854	26	118	4	98.495	7
112	5	21163.600	45	118	5	119.188	3
112	6	19070.066	31	118	6	94.799	2
112	7	18809.666	45	118	7	70.064	3
113	1	2582.738	18	119	1	5100.537	23
113	2	3288.781	25	119	2	5307.799	26
113	3	3107.033	25	119	3	5885.003	31
113	4	3201.166	17	119	4	6305.052	22
113	5	4361.026	46	119	5	8304.123	41
113	6	3310.561	33	119	6	7028.993	46
113	7	2020.386	12	119	7	6541.694	37
114	1	3178.428	3	120	1	1310.079	24
114	2	2294.945	11	120	2	1596.148	38
114	3	2563.539	8	120	3	1577.690	34
114	4	2828.968	6	120	4	1671.204	38
114	5	4422.089	9	120	5	2328.100	73
114	6	3809.180	14	120	6	2066.956	63
114	7	4112.043	9	120	7	1728.151	50

Data Set 2

Class	Year	Payroll	Loss	Class	Year	Payroll	Loss
121	1	66.772	0	127	1	180.970	4
121	2	70.669	0	127	2	249.116	5
121	3	70.737	0	127	3	221.405	7
121	4	73.274	0	127	4	161.425	7
121	5	87.124	2	127	5	199.852	6
121	6	83.261	0	127	6	172.033	6
121	7	89.242	0	127	7	140.378	7
122	1	2113.111	17	128	1	0.000	0
122	2	2294.597	38	128	2	0.000	0
122	3	2585.560	38	128	3	0.000	0
122	4	2524.492	53	128	4	0.000	0
122	5	3616.607	78	128	5	0.000	0
122	6	2887.929	81	128	6	0.000	0
122	7	2870.043	67	128	7	0.000	0
123	1	104.253	2	129	1	795.167	19
123	2	138.179	2	129	2	1147.475	26
123	3	139.056	2	129	3	950.837	27
123	4	172.297	5	129	4	849.241	15
123	5	187.190	5	129	5	976.607	22
123	6	132.517	5	129	6	673.329	16
123	7	142.657	2	129	7	535.045	15
124	1	3.275	0	130	1	82.610	0
124	2	5.693	0	130	2	87.124	2
124	3	4.508	0	130	3	111.903	0
124	4	4.805	0	130	4	123.734	1
124	5	9.067	0	130	5	120.606	3
124	6	5.967	0	130	6	148.423	1
124	7	5.588	0	130	7	70.599	0
125	1	13.182	3	131	1	61.802	0
125	2	20.734	2	131	2	116.093	2
125	3	15.334	3	131	3	64.425	2
125	4	17.762	1	131	4	56.817	0
125	5	17.228	0	131	5	92.627	1
125	6	1.666	11	131	6	119.006	6
125	7	1.740	1	131	7	120.145	8
126	1	76.205	1	132	1	2.317	0
126	2	70.900	0	132	2	3.972	0
126	3	87.235	2	132	3	33.190	0
126	4	100.693	0	132	4	47.261	2
126	5	143.081	0	132	5	35.413	2
126	6	120.950	0	132	6	16.984	1
126	7	123.444	2	132	7	18.797	0

Data Set 2

Class	Year	Payroll	Loss
133	1	66.213	2
133	2	65.696	0
133	3	69.913	0
133	4	46.289	0
133	5	91.119	0
133	6	78.456	1
133	7	84.362	2

Data Set 3

Cl	Yr	Payroll	Loss	Cl	Yr	Payroll	Loss
1	1	21798086	538707	8	1	3724269	0
1	2	22640528	439184	8	2	4149906	18082
1	3	22572010	1059775	8	3	3696872	408802
1	4	24789710	560013	8	4	487892	10500
1	5	25876764	1004997	8	5	594629	0
1	6	28033613	1097314	8	6	560057	0
1	7	22525887	609833	8	7	724571	37500
2	1	12004031	270222	9	1	2494260	11234
2	2	12713178	229566	9	2	3394857	42146
2	3	13596610	596850	9	3	3314038	86599
2	4	14811727	196539	9	4	10595191	41500
2	5	12774073	134248	9	5	12863955	115592
2	6	20245789	489312	9	6	2789484	0
2	7	24242468	418218	9	7	2426480	170623
3	1	50216515	769208	10	1	3140245	78436
3	2	56099793	649707	10	2	3964679	47501
3	3	58109747	503919	10	3	4899610	51499
3	4	67807105	675466	10	4	5247896	144621
3	5	73852437	545745	10	5	6280799	8998
3	6	84208474	1562266	10	6	7721616	259875
3	7	83604216	931762	10	7	9324806	548369
4	1	19507341	291125	11	1	30035148	383658
4	2	20810822	499927	11	2	32618878	420764
4	3	23838122	151084	11	3	36071278	588094
4	4	24212193	141910	11	4	42494763	717523
4	5	25367578	300373	11	5	49524793	1133547
4	6	35401344	119761	11	6	48872610	1372065
4	7	37580989	141300	11	7	40636397	601273
5	1	12428727	377662	12	1	211719199	2140846
5	2	13492474	169574	12	2	232681695	2382040
5	3	13303600	243585	12	3	230007534	2929906
5	4	13228920	61215	12	4	234635378	3165114
5	5	14060473	55272	12	5	251852677	3595126
5	6	15227696	312816	12	6	207512194	4284873
5	7	17857683	160196	12	7	215038202	2739517
6	1	6829432	623029	13	1	22525133	389158
6	2	6219655	280707	13	2	23545828	374650
6	3	5696577	174422	13	3	24979337	684493
6	4	4542884	176666	13	4	27390359	143933
6	5	3663628	153907	13	5	27337832	1236701
6	6	2632086	101874	13	6	28268401	734178
6	7	1932376	0	13	7	25281128	850133

Data Set 3

Cl	Yr	Payroll	Loss	Cl	Yr	Payroll	Loss
14	1	11939588	348156	20	1	251151	0
14	2	10566079	331977	20	2	265961	0
14	3	19387104	116060	20	3	584921	39110
14	4	23062327	266805	20	4	912410	0
14	5	23089131	310604	20	5	790496	182828
14	6	23799825	278189	20	6	1459268	158752
14	7	21317594	79680	20	7	1198147	15218
15	1	28171269	1238292	21	1	1066125	0
15	2	32008767	576512	21	2	1248303	76795
15	3	31300733	831404	21	3	1685603	0
15	4	33507831	1303705	21	4	955740	0
15	5	32207902	673907	21	5	2168747	36088
15	6	23471988	605122	21	6	1843376	62788
15	7	24101833	664457	21	7	2353645	69600
16	1	24766193	348918	22	1	4721960	126583
16	2	26078216	370878	22	2	4936589	124071
16	3	30044642	679015	22	3	5988667	112488
16	4	31886024	815234	22	4	6836759	158834
16	5	37129722	787158	22	5	7670467	532500
16	6	35982441	867186	22	6	5103933	43724
16	7	35300899	434384	22	7	4365821	3500
17	1	38056609	501744	23	1	907636	0
17	2	27027036	490322	23	2	1145044	0
17	3	34508011	288718	23	3	1278710	0
17	4	47219834	733864	23	4	1248136	0
17	5	47544814	1089948	23	5	1311819	0
17	6	38168079	663617	23	6	1362253	0
17	7	32437577	582524	23	7	956184	0
18	1	24859345	111484	25	1	23173030	117766
18	2	31348102	292036	25	2	27970610	243929
18	3	34311081	309761	25	3	32939613	352193
18	4	38116969	903327	25	4	34898819	576683
18	5	40921687	309404	25	5	38666318	203312
18	6	46388415	844651	25	6	39018294	639064
18	7	29413879	622879	25	7	43386511	423313
19	1	6625	0	26	1	8155680	368612
19	2	11130	0	26	2	10010676	227886
19	3	276172	0	26	3	6232545	224393
19	4	116042	0	26	4	7394027	269011
19	5	13727	0	26	5	8827295	187958
19	6	11289	0	26	6	8244511	270941
19	7	7509	0	26	7	8357388	103073

Data Set 3

Cl	Yr	Payroll	Loss	Cl	Yr	Payroll	Loss
27	1	33174216	992111	33	1	15534295	147874
27	2	33929034	861857	33	2	17652385	189352
27	3	33930095	1229651	33	3	19034096	157894
27	4	31335349	1194090	33	4	15078778	219586
27	5	27986197	674696	33	5	19666426	499935
27	6	27960360	911083	33	6	9352445	136531
27	7	24171183	595925	33	7	11897928	107634
28	1	24864691	171768	34	1	38033198	401014
28	2	24513649	478225	34	2	47911424	143187
28	3	24202768	304261	34	3	55326918	183812
28	4	29462623	405819	34	4	66707182	247650
28	5	30570568	490959	34	5	77034529	222286
28	6	32674597	655637	34	6	75453903	217292
28	7	33601901	194720	34	7	76011263	108300
29	1	21143095	193805	35	1	24679406	20496
29	2	21949119	591933	35	2	29243897	52714
29	3	33006399	1376195	35	3	33420158	83450
29	4	35012750	1569869	35	4	26693266	9500
29	5	35093028	1770362	35	5	23889630	10642
29	6	31437635	1584526	35	6	24554121	149565
29	7	38785568	686330	35	7	16712481	0
30	1	14063006	83244	36	1	81639041	2315716
30	2	17535057	97882	36	2	88096749	2068151
30	3	17521388	230937	36	3	96589758	2348199
30	4	20951695	166281	36	4	90214474	2927326
30	5	22305798	80926	36	5	91415878	2076388
30	6	23607400	145161	36	6	78948454	2192662
30	7	25766947	251852	36	7	75541403	1315745
31	1	31844283	727509	37	1	141679901	1039667
31	2	33844240	637332	37	2	152507292	1168176
31	3	43055585	824169	37	3	151544121	1684628
31	4	46934412	1212279	37	4	172714585	1709070
31	5	57825009	1092508	37	5	188112741	2241146
31	6	43149035	1058647	37	6	188540013	1846560
31	7	40664379	448471	37	7	188080205	795193
32	1	109890415	2108179	38	1	138433273	2270992
32	2	138889855	4121925	38	2	148751843	2670269
32	3	172495213	3687446	38	3	171364057	3154485
32	4	184236306	3968537	38	4	180969565	2910129
32	5	198716792	5204079	38	5	190962808	4194534
32	6	161794404	3881944	38	6	163181576	2940649
32	7	154990570	2142559	38	7	144772872	1190392

Data Set 3

Cl	Yr	Payroll	Loss	Cl	Yr	Payroll	Loss
39	1	7584291	188300	45	1	1076853336	14791523
39	2	7663938	216447	45	2	1160411930	10323164
39	3	5816169	227945	45	3	1358874727	15691903
39	4	4519795	120519	45	4	1477578301	16481920
39	5	4173473	239610	45	5	1608439866	22540025
39	6	2813954	74187	45	6	1614075733	18924240
39	7	1751035	92809	45	7	1587379829	12302933
40	1	1679525	257004	46	1	400776887	13167187
40	2	1659623	22500	46	2	439405026	11588936
40	3	2100761	179596	46	3	515789316	13164872
40	4	2444614	78045	46	4	511546484	17560626
40	5	2290289	136500	46	5	508739866	21897988
40	6	1769603	132421	46	6	417518604	13886374
40	7	16939	0	46	7	412916221	9281573
41	1	89617108	1824667	47	1	9124894	104514
41	2	105279332	1354204	47	2	8640154	64351
41	3	121675468	1615600	47	3	7561489	220480
41	4	113759934	3439832	47	4	13351122	260506
41	5	106241178	2372270	47	5	17648304	149896
41	6	117696551	2781589	47	6	13835781	177491
41	7	131918966	1145657	47	7	15799659	151138
42	1	39189024	1774180	48	1	14227234	225349
42	2	40208840	828369	48	2	14662758	372707
42	3	42603592	1164966	48	3	13563646	266912
42	4	41303042	1312461	48	4	16049163	180843
42	5	46864132	939836	48	5	7222573	141104
42	6	49639346	2042816	48	6	6470960	464746
42	7	26858912	553718	48	7	7127968	51500
43	1	493502986	6930096	49	1	72009717	471987
43	2	523950857	6148910	49	2	81190936	382908
43	3	596534761	7711975	49	3	83294158	678766
43	4	648381669	9269550	49	4	101533560	928766
43	5	756980175	12033862	49	5	114551541	1873704
43	6	717355951	8425861	49	6	120071617	807600
43	7	671740789	6130321	49	7	127493896	320233
44	1	188580465	3181329	50	1	21790188	1084102
44	2	230254574	3396889	50	2	33262400	506559
44	3	258723935	4883645	50	3	31263030	673649
44	4	274052331	6288206	50	4	40355211	779146
44	5	306701321	5611726	50	5	35877383	857853
44	6	252446006	4153754	50	6	22915935	358744
44	7	228485713	2285936	50	7	19895890	376492

Data Set 3

Cl	Yr	Payroll	Loss	Cl	Yr	Payroll	Loss
51	1	757821	0	58	1	0	0
51	2	757641	0	58	2	2060821	0
51	3	7335637	8431	58	3	450607	26867
51	4	7383931	8463	58	4	3407286	0
51	5	6786271	3686	58	5	1400342	0
51	6	656806	0	58	6	0	0
51	7	1043359	0	58	7	1856138	0
52	1	5029937	10887	59	1	39375739	183729
52	2	6405035	103272	59	2	41660941	253380
52	3	8962965	72445	59	3	41434679	343244
52	4	6565558	38735	59	4	45915594	225532
52	5	8298029	116024	59	5	48077645	105553
52	6	8299929	283258	59	6	46190580	250581
52	7	7765637	105850	59	7	48001269	60450
53	1	23104764	175142	60	1	163342957	773290
53	2	27125181	261911	60	2	88046040	950805
53	3	29397033	330441	60	3	209330794	2318372
53	4	34084793	220702	60	4	223405673	1966799
53	5	34811381	221741	60	5	115906575	2083123
53	6	42012182	619084	60	6	47200765	2481152
53	7	34210076	364423	60	7	37125028	516619
55	1	3016725	98664	61	1	1582136	33950
55	2	2110339	82792	61	2	404018	0
55	3	2978128	45235	61	3	2418832	0
55	4	4947120	129101	61	4	2572060	18412
55	5	5245604	239601	61	5	40020	0
55	6	3941599	156914	61	6	91290	0
55	7	2935881	32762	61	7	151329	0
56	1	12111455	315162	62	1	20468845	302620
56	2	15593028	175347	62	2	23377052	726287
56	3	20867120	232339	62	3	26147695	292036
56	4	21410061	148191	62	4	25801216	248576
56	5	20919053	148971	62	5	36527190	1001047
56	6	22824203	234977	62	6	33681736	142442
56	7	7992500	35120	62	7	26612618	88206
57	1	22794762	377559	63	1	24779341	186710
57	2	16358650	225066	63	2	26577845	108013
57	3	21397233	201408	63	3	25679916	166504
57	4	24139786	258598	63	4	33945076	168370
57	5	25935295	376629	63	5	32490822	477428
57	6	26318495	583853	63	6	66772790	334805
57	7	21513579	154282	63	7	8337246	50451

Data Set 3

Cl	Yr	Payroll	Loss	Cl	Yr	Payroll	Loss
64	1	67597823	350694	70	1	20012771	86790
64	2	72333674	185068	70	2	22555092	54500
64	3	77305484	131137	70	3	22996619	0
64	4	86131464	477613	70	4	23139219	40874
64	5	89236341	132406	70	5	26508048	185322
64	6	116875583	305702	70	6	26721238	88744
64	7	116536546	169213	70	7	26143975	32509
65	1	27125589	289393	71	1	30697342	325094
65	2	31026734	520115	71	2	33505572	730330
65	3	31952471	733091	71	3	39007165	804643
65	4	38678601	971153	71	4	45700829	656593
65	5	50135420	552827	71	5	49336957	960338
65	6	51349900	432620	71	6	45958189	839480
65	7	46002657	141228	71	7	43258125	819031
66	1	5490003	46512	72	1	28237602	559718
66	2	5130725	33179	72	2	32931312	301679
66	3	6346136	57335	72	3	33227406	342556
66	4	7830414	79260	72	4	39383623	459514
66	5	9003930	57728	72	5	39710664	369830
66	6	4059519	15671	72	6	32135732	463212
66	7	4345173	16932	72	7	29886290	138668
67	1	12937143	53238	73	1	752034	0
67	2	13812853	222145	73	2	857456	17000
67	3	14600120	152171	73	3	798471	0
67	4	16720545	405253	73	4	468790	0
67	5	18259745	895203	73	5	609284	21563
67	6	11611818	325950	73	6	499244	2550
67	7	12050112	259307	73	7	341978	7600
68	1	68136	0	74	1	149522987	2243821
68	2	228763	0	74	2	155921438	2935172
68	3	316733	0	74	3	149463133	2551411
68	4	266400	0	74	4	156495587	2541658
68	5	153959	0	74	5	132706576	3491156
68	6	94486	0	74	6	104588639	2741726
68	7	158993	0	74	7	79418153	1781270
69	1	3372342	12618	75	1	25969226	936384
69	2	4957866	16738	75	2	33107562	771134
69	3	4976225	79200	75	3	37053424	1176644
69	4	6110830	62141	75	4	40105604	2238470
69	5	7063073	49300	75	5	42137626	1705827
69	6	5378775	30529	75	6	44317286	2402206
69	7	5405090	18750	75	7	42831726	998274

Data Set 3

Cl	Yr	Payroll	Loss	Cl	Yr	Payroll	Loss
76	1	109727515	2505721	82	1	484542478	5873364
76	2	111250557	2600133	82	2	516410679	5580000
76	3	114566586	2366325	82	3	612718970	6580660
76	4	143133426	2936005	82	4	778141048	8480126
76	5	179852656	4980010	82	5	933927586	11747286
76	6	158632827	4554005	82	6	837460151	11506309
76	7	130277681	2170374	82	7	823540456	6363269
77	1	24657446	428505	83	1	337016621	8341391
77	2	27252047	584198	83	2	383030593	7947246
77	3	34313427	780737	83	3	444858785	9957382
77	4	48713355	568411	83	4	494127462	10696279
77	5	62463163	996150	83	5	573459505	14952185
77	6	62821338	1851949	83	6	460471568	11975633
77	7	58167406	1437342	83	7	411094525	8088591
78	1	4635081	71932	84	1	1678351	268983
78	2	6676915	76044	84	2	2766610	293717
78	3	7163162	1040582	84	3	683652	0
78	4	7509642	278890	84	4	572365	0
78	5	6091676	297011	84	5	1233345	0
78	6	4801644	16680	84	6	790382	0
78	7	3798499	19570	84	7	2410747	46983
79	1	26941444	1523050	85	1	19139111	146591
79	2	27005374	726021	85	2	15250014	450069
79	3	40925748	2227595	85	3	5937661	120038
79	4	42794092	2241043	85	4	5748909	54334
79	5	50533403	2686884	85	5	6099707	79225
79	6	42853732	1235643	85	6	6393692	104842
79	7	43151550	1339032	85	7	6396749	54913
80	1	241242170	4064107	86	1	233478	0
80	2	260050016	3507860	86	2	7248724	38297
80	3	293874853	5673842	86	3	20372214	380450
80	4	347610263	6600927	86	4	24438753	599656
80	5	407206410	8076512	86	5	30284192	733591
80	6	381240905	9048778	86	6	25440184	541206
80	7	316827641	4835817	86	7	19053737	216545
81	1	21834276	1116947	87	1	2555760	3999
81	2	24289389	1156293	87	2	3793129	0
81	3	26590185	947665	87	3	3516163	0
81	4	33085882	1131602	87	4	4147374	20000
81	5	37393442	1138077	87	5	5415438	10200
81	6	37085072	1942467	87	6	3949638	0
81	7	29484404	1769216	87	7	4637883	0

Data Set 3

Cl	Yr	Payroll	Loss	Cl	Yr	Payroll	Loss
88	1	1882882	491155	94	1	36903323	350667
88	2	595706	0	94	2	40984574	306021
88	3	699763	9500	94	3	39119635	497597
88	4	2642635	41239	94	4	34256426	208057
88	5	880862	0	94	5	30784622	526764
88	6	1031491	0	94	6	27546781	457679
88	7	571529	0	94	7	25519552	159473
89	1	433689	23110	95	1	76861184	543694
89	2	579295	52965	95	2	84860361	559360
89	3	1888551	129363	95	3	97459955	439502
89	4	1062573	72931	95	4	91556250	302649
89	5	764529	119664	95	5	104160964	931385
89	6	575091	212263	95	6	108159477	641728
89	7	518852	32000	95	7	122262246	547258
90	1	992135	17446	96	1	22455361	574249
90	2	1018086	18753	96	2	24508893	696819
90	3	899186	0	96	3	24573769	672280
90	4	1145576	96760	96	4	24264175	513825
90	5	3892916	286605	96	5	22564774	533450
90	6	2179505	29287	96	6	21114872	697521
90	7	1374750	0	96	7	19292975	757987
91	1	286652537	2468560	97	1	75482127	864694
91	2	327549794	3401366	97	2	83118489	839390
91	3	363596957	5422494	97	3	88285454	908835
91	4	384610394	6009402	97	4	101112977	1088671
91	5	410095260	7682568	97	5	110779799	2269138
91	6	425206918	6919518	97	6	102279747	1852660
91	7	409026813	5092578	97	7	90206335	1118383
92	1	250065709	4785125	98	1	653956492	4689667
92	2	303706645	6422628	98	2	590077975	4001950
92	3	315668497	7722638	98	3	563942403	5410725
92	4	395667661	10047060	98	4	671763488	3635426
92	5	394518644	11384364	98	5	721014971	7149183
92	6	349603983	9324049	98	6	747547369	5500193
92	7	319726088	6369069	98	7	675097746	4281574
93	1	60318819	121210	99	1	192750431	2057381
93	2	74169286	32054	99	2	211141349	2732913
93	3	89774403	12235	99	3	229248184	2697519
93	4	102539406	58805	99	4	245337377	4117435
93	5	130862582	84184	99	5	287634618	4119698
93	6	157126568	696627	99	6	293068300	4022960
93	7	163636648	247033	99	7	273133614	3032105

Data Set 3

Cl	Yr	Payroll	Loss	Cl	Yr	Payroll	Loss
100	1	21709931	177713	106	1	4147121	13100
100	2	23832098	174695	106	2	4538955	0
100	3	25989455	155791	106	3	4790566	222410
100	4	28210859	458737	106	4	4937697	76865
100	5	33382788	372560	106	5	6350938	29800
100	6	35185420	260230	106	6	5599943	346534
100	7	38657390	118634	106	7	5245733	52363
101	1	477018232	1916583	107	1	48764143	21015
101	2	537842107	2218605	107	2	54368435	26185
101	3	586805062	2382946	107	3	70495113	60000
101	4	647922253	2360265	107	4	83575874	48904
101	5	730285763	5256806	107	5	94560530	97347
101	6	784741752	4312805	107	6	110016620	77243
101	7	748156291	2634603	107	7	115686229	289763
102	1	31405266	478173	108	1	220681335	2680864
102	2	36263323	872631	108	2	237076525	3533857
102	3	39662202	672755	108	3	263297154	4362730
102	4	45620069	920983	108	4	284208600	5659167
102	5	49803495	510466	108	5	299761668	6522245
102	6	48965697	739138	108	6	290714989	6439913
102	7	44321882	914435	108	7	229984443	5159319
103	1	59054933	1738019	109	1	18690811	553599
103	2	72098358	2358683	109	2	21991832	686131
103	3	79607959	2679043	109	3	25431630	373331
103	4	80316693	3855668	109	4	29757498	621559
103	5	90814765	4293751	109	5	29583030	930449
103	6	76619602	2424582	109	6	31972861	625163
103	7	72978774	1418296	109	7	29803131	303244
104	1	8678921	346227	110	1	18370543	421173
104	2	9803526	354702	110	2	19169416	476994
104	3	8784862	181314	110	3	20941825	200669
104	4	9394166	400532	110	4	21343008	155689
104	5	10068619	114668	110	5	21285402	329873
104	6	9650435	156866	110	6	23113364	300760
104	7	8764544	198038	110	7	24134032	183234
105	1	514054148	5610971	111	1	14092284	444330
105	2	557293311	5608315	111	2	13612973	551408
105	3	606081740	5094525	111	3	14699823	609969
105	4	291197602	3413845	111	4	17100322	736448
105	5	302054660	4433195	111	5	18024042	781660
105	6	283185968	3833388	111	6	15469163	651095
105	7	272699866	3606637	111	7	14063564	731923

Data Set 3

Cl	Yr	Payroll	Loss	Cl	Yr	Payroll	Loss
112	1	3416861519	2052161	118	1	22640141	353000
112	2	3784782293	2779991	118	2	26982541	300183
112	3	4284555751	2814303	118	3	25879842	514545
112	4	4866862481	4196086	118	4	32376841	556276
112	5	5559891691	6202670	118	5	31025023	426913
112	6	5948227717	5357248	118	6	34728096	682990
112	7	6137275140	6633541	118	7	28692495	69744
113	1	690442346	2841730	119	1	1223456875	2018392
113	2	630445853	3072933	119	2	1434190104	1698731
113	3	470504472	2469680	119	3	1621261681	2602445
113	4	462726271	1737875	119	4	1976427753	2691596
113	5	347548389	1698106	119	5	2240090280	4959226
113	6	269458657	1457243	119	6	2257188913	5748591
113	7	302096838	706363	119	7	2142352263	3234049
114	1	1773712734	1912434	120	1	480721919	5930220
114	2	2169653363	1863796	120	2	489791825	6883545
114	3	2253443315	1708796	120	3	468110661	7867316
114	4	2262445806	2437372	120	4	477107053	8691719
114	5	2140274698	2723582	120	5	435089059	8643069
114	6	1892565179	2069504	120	6	398898500	7134252
114	7	1886357964	1014231	120	7	412578720	6249488
115	1	61527503	384495	121	1	17532972	6961
115	2	57870396	593615	121	2	19607417	83190
115	3	70838124	266751	121	3	22462641	98328
115	4	76543560	400315	121	4	22253695	101140
115	5	91358603	1085454	121	5	24729365	148597
115	6	98839324	471234	121	6	29264458	128268
115	7	95944888	418734	121	7	28043076	112138
116	1	142604414	567583	122	1	688447555	4742669
116	2	153758493	512948	122	2	821766634	5096536
116	3	187642951	1419591	122	3	830349943	6008706
116	4	211609452	895304	122	4	955695913	7061399
116	5	258710007	819245	122	5	1048747380	9200630
116	6	257181827	1231375	122	6	1080858674	10075347
116	7	272490770	733850	122	7	1053140929	6244481
117	1	58773481	956162	123	1	36052962	205307
117	2	58952303	1212906	123	2	40316407	340822
117	3	62416901	885291	123	3	39483537	158653
117	4	65321406	576135	123	4	42927612	128628
117	5	55164886	632216	123	5	44164598	143564
117	6	32777500	916972	123	6	37715617	461927
117	7	30099988	431813	123	7	34613171	375239

Data Set 3

Cl	Yr	Payroll	Loss
124	1	3788569	57571
124	2	4216515	46100
124	3	5397424	150931
124	4	5585309	196756
124	5	5623680	367050
124	6	4792099	229725
124	7	3544705	161360

Data Set 4

State	Occ	Yr	Exp.	Loss	State	Occ	Yr	Exp.	Loss
1	1	1	32.322	1	1	7	1	0.000	0
1	1	2	33.779	4	1	7	2	0.000	0
1	1	3	43.548	3	1	7	3	0.000	0
1	1	4	46.686	5	1	7	4	0.000	0
1	1	5	34.713	1	1	7	5	0.000	0
1	1	6	32.857	3	1	7	6	0.000	0
1	1	7	36.600	4	1	7	7	0.000	0
1	2	1	45.995	3	1	8	1	41.403	3
1	2	2	37.888	1	1	8	2	34.066	11
1	2	3	34.581	0	1	8	3	32.729	4
1	2	4	28.298	0	1	8	4	32.235	0
1	2	5	45.265	2	1	8	5	32.777	2
1	2	6	39.945	0	1	8	6	28.620	4
1	2	7	39.322	4	1	8	7	24.263	1
1	3	1	289.047	5	1	9	1	6.452	0
1	3	2	392.176	8	1	9	2	6.927	1
1	3	3	368.982	8	1	9	3	5.851	0
1	3	4	323.770	8	1	9	4	3.033	1
1	3	5	385.222	16	1	9	5	1.787	0
1	3	6	346.390	8	1	9	6	1.074	1
1	3	7	324.132	9	1	9	7	0.685	0
1	4	1	0.000	0	1	10	1	537.311	11
1	4	2	0.000	0	1	10	2	569.041	12
1	4	3	0.000	0	1	10	3	597.146	15
1	4	4	0.037	0	1	10	4	570.295	15
1	4	5	0.000	0	1	10	5	1116.750	18
1	4	6	0.000	0	1	10	6	774.454	10
1	4	7	0.000	0	1	10	7	534.953	11
1	5	1	310.389	15	1	11	1	149.683	6
1	5	2	292.464	10	1	11	2	157.947	6
1	5	3	262.560	18	1	11	3	174.549	5
1	5	4	273.257	25	1	11	4	181.317	10
1	5	5	372.730	27	1	11	5	202.066	13
1	5	6	263.443	20	1	11	6	187.564	7
1	5	7	355.826	28	1	11	7	229.830	8
1	6	1	291.784	27	1	12	1	609.467	20
1	6	2	267.439	44	1	12	2	645.375	20
1	6	3	288.555	40	1	12	3	667.384	27
1	6	4	308.364	31	1	12	4	573.144	15
1	6	5	387.095	40	1	12	5	782.643	38
1	6	6	351.445	34	1	12	6	478.749	8
1	6	7	342.518	37	1	12	7	454.967	13

Data Set 4

State	Occ	Yr	Exp.	Loss	State	Occ	Yr	Exp.	Loss
1	13	1	120.027	0	1	19	1	77.909	3
1	13	2	131.020	1	1	19	2	78.925	0
1	13	3	161.145	8	1	19	3	77.558	1
1	13	4	182.135	5	1	19	4	69.525	1
1	13	5	276.520	11	1	19	5	65.972	0
1	13	6	158.310	2	1	19	6	63.274	0
1	13	7	168.420	7	1	19	7	75.335	2
1	14	1	196.722	13	1	20	1	1787.463	50
1	14	2	209.923	12	1	20	2	2027.230	90
1	14	3	196.199	13	1	20	3	1853.812	54
1	14	4	188.820	14	1	20	4	1742.135	63
1	14	5	202.807	19	1	20	5	2119.499	56
1	14	6	180.979	8	1	20	6	1545.169	35
1	14	7	195.628	6	1	20	7	1315.368	22
1	15	1	215.898	11	1	21	1	258.885	10
1	15	2	224.229	11	1	21	2	314.825	10
1	15	3	224.306	10	1	21	3	299.999	10
1	15	4	212.232	6	1	21	4	314.081	11
1	15	5	245.198	11	1	21	5	258.487	17
1	15	6	203.698	16	1	21	6	235.082	8
1	15	7	210.496	15	1	21	7	230.142	8
1	16	1	57.265	4	1	22	1	86.381	4
1	16	2	62.662	3	1	22	2	95.036	4
1	16	3	66.984	6	1	22	3	134.285	4
1	16	4	63.211	3	1	22	4	113.491	8
1	16	5	77.680	4	1	22	5	141.717	0
1	16	6	61.944	3	1	22	6	128.956	0
1	16	7	44.195	3	1	22	7	105.489	1
1	17	1	347.835	22	1	23	1	766.953	7
1	17	2	326.396	7	1	23	2	912.516	8
1	17	3	307.978	13	1	23	3	866.413	9
1	17	4	350.914	20	1	23	4	898.687	15
1	17	5	546.047	23	1	23	5	1806.752	34
1	17	6	410.980	8	1	23	6	1018.684	14
1	17	7	377.287	18	1	23	7	984.843	11
1	18	1	0.000	0	1	24	1	62.153	1
1	18	2	0.000	0	1	24	2	84.116	1
1	18	3	0.000	0	1	24	3	75.524	1
1	18	4	0.000	0	1	24	4	109.237	1
1	18	5	0.000	0	1	24	5	119.034	2
1	18	6	0.000	0	1	24	6	116.794	1
1	18	7	0.000	0	1	24	7	136.571	3

Data Set 4

State	Occ	Yr	Exp.	Loss	State	Occ	Yr	Exp.	Loss
1	25	1	3431.494	59	2	6	1	213.491	106
1	25	2	3882.069	68	2	6	2	236.267	102
1	25	3	3805.563	76	2	6	3	237.557	95
1	25	4	3919.527	77	2	6	4	240.764	84
1	25	5	4352.809	94	2	6	5	302.906	114
1	25	6	3949.550	73	2	6	6	238.717	130
1	25	7	3927.784	73	2	6	7	235.381	80
2	1	1	26.813	0	2	7	1	0.000	0
2	1	2	25.941	2	2	7	2	0.000	0
2	1	3	24.812	3	2	7	3	0.000	0
2	1	4	33.346	1	2	7	4	0.000	0
2	1	5	38.921	3	2	7	5	0.000	0
2	1	6	34.110	5	2	7	6	0.000	0
2	1	7	32.772	11	2	7	7	0.000	0
2	2	1	7.445	0	2	8	1	130.502	14
2	2	2	32.768	5	2	8	2	100.659	7
2	2	3	40.518	5	2	8	3	112.432	11
2	2	4	43.832	1	2	8	4	101.994	17
2	2	5	52.982	8	2	8	5	93.142	17
2	2	6	43.574	6	2	8	6	69.517	10
2	2	7	42.812	4	2	8	7	54.891	8
2	3	1	107.791	5	2	9	1	12.684	3
2	3	2	110.400	10	2	9	2	6.981	2
2	3	3	109.904	10	2	9	3	1.882	0
2	3	4	110.862	8	2	9	4	4.855	1
2	3	5	144.156	10	2	9	5	4.591	0
2	3	6	109.533	8	2	9	6	2.908	0
2	3	7	114.972	12	2	9	7	2.347	1
2	4	1	0.000	0	2	10	1	164.629	4
2	4	2	1.591	0	2	10	2	248.315	4
2	4	3	2.926	0	2	10	3	187.910	12
2	4	4	3.387	0	2	10	4	181.905	12
2	4	5	3.559	0	2	10	5	225.843	8
2	4	6	4.419	0	2	10	6	227.746	7
2	4	7	1.422	0	2	10	7	194.361	12
2	5	1	501.176	60	2	11	1	143.854	17
2	5	2	439.284	39	2	11	2	132.329	19
2	5	3	434.990	31	2	11	3	132.716	10
2	5	4	479.366	24	2	11	4	140.268	14
2	5	5	547.592	54	2	11	5	171.932	17
2	5	6	419.615	41	2	11	6	150.514	11
2	5	7	393.227	45	2	11	7	105.342	5

Data Set 4

State	Occ	Yr	Exp.	Loss	State	Occ	Yr	Exp.	Loss
2	12	1	520.104	29	2	18	1	0.000	0
2	12	2	528.886	29	2	18	2	0.000	0
2	12	3	504.641	23	2	18	3	0.000	0
2	12	4	631.105	22	2	18	4	0.000	0
2	12	5	687.263	38	2	18	5	0.000	0
2	12	6	520.756	30	2	18	6	0.000	0
2	12	7	505.274	16	2	18	7	2.053	0
2	13	1	46.957	0	2	19	1	0.000	0
2	13	2	46.131	2	2	19	2	0.000	0
2	13	3	52.492	4	2	19	3	0.000	0
2	13	4	67.550	7	2	19	4	0.000	0
2	13	5	89.129	10	2	19	5	0.000	0
2	13	6	92.599	19	2	19	6	0.000	0
2	13	7	83.307	27	2	19	7	0.000	0
2	14	1	208.720	40	2	20	1	157.453	11
2	14	2	200.651	17	2	20	2	188.484	20
2	14	3	213.847	26	2	20	3	193.104	8
2	14	4	240.331	38	2	20	4	184.118	10
2	14	5	317.311	30	2	20	5	217.162	7
2	14	6	261.859	31	2	20	6	183.674	5
2	14	7	262.810	22	2	20	7	209.780	8
2	15	1	165.747	9	2	21	1	0.000	0
2	15	2	122.684	13	2	21	2	0.107	0
2	15	3	139.453	18	2	21	3	0.000	0
2	15	4	171.691	12	2	21	4	0.917	0
2	15	5	121.177	12	2	21	5	0.392	0
2	15	6	134.710	28	2	21	6	1.105	0
2	15	7	161.964	31	2	21	7	0.000	0
2	16	1	152.678	3	2	22	1	123.466	3
2	16	2	168.215	8	2	22	2	145.501	6
2	16	3	121.978	13	2	22	3	152.616	8
2	16	4	112.014	11	2	22	4	155.315	5
2	16	5	180.867	13	2	22	5	165.100	6
2	16	6	98.333	4	2	22	6	133.533	6
2	16	7	96.711	3	2	22	7	132.168	4
2	17	1	213.770	4	2	23	1	83.347	0
2	17	2	225.099	22	2	23	2	46.994	0
2	17	3	245.192	15	2	23	3	34.547	0
2	17	4	248.563	16	2	23	4	23.462	0
2	17	5	262.016	14	2	23	5	24.235	0
2	17	6	247.521	23	2	23	6	0.648	0
2	17	7	236.203	9	2	23	7	5.159	0

Data Set 4

State	Occ	Yr	Exp.	Loss	State	Occ	Yr	Exp.	Loss
2	24	1	42.786	3	3	5	1	0.000	0
2	24	2	58.532	10	3	5	2	0.287	0
2	24	3	64.672	2	3	5	3	0.000	0
2	24	4	57.346	4	3	5	4	0.671	0
2	24	5	63.072	3	3	5	5	0.347	0
2	24	6	45.723	5	3	5	6	0.414	0
2	24	7	54.513	0	3	5	7	0.000	0
2	25	1	944.298	40	3	6	1	0.000	0
2	25	2	1101.740	59	3	6	2	0.000	0
2	25	3	1117.233	35	3	6	3	0.000	0
2	25	4	1006.895	51	3	6	4	0.000	0
2	25	5	1100.779	27	3	6	5	0.000	0
2	25	6	825.368	24	3	6	6	0.000	0
2	25	7	680.958	9	3	6	7	0.000	0
3	1	1	2.220	1	3	7	1	0.000	0
3	1	2	1.920	0	3	7	2	0.000	0
3	1	3	2.229	0	3	7	3	0.000	0
3	1	4	2.325	0	3	7	4	0.000	0
3	1	5	1.815	0	3	7	5	0.000	0
3	1	6	2.149	1	3	7	6	0.000	0
3	1	7	2.292	1	3	7	7	0.000	0
3	2	1	0.000	0	3	8	1	0.000	0
3	2	2	0.006	0	3	8	2	0.000	0
3	2	3	0.000	0	3	8	3	0.000	0
3	2	4	0.000	0	3	8	4	0.000	0
3	2	5	0.000	0	3	8	5	0.000	0
3	2	6	0.000	0	3	8	6	0.000	0
3	2	7	0.000	0	3	8	7	0.000	0
3	3	1	9.515	0	3	9	1	0.000	0
3	3	2	19.685	1	3	9	2	0.000	0
3	3	3	21.632	2	3	9	3	0.000	0
3	3	4	24.686	0	3	9	4	0.000	0
3	3	5	15.567	2	3	9	5	0.000	0
3	3	6	19.270	2	3	9	6	0.000	0
3	3	7	17.241	3	3	9	7	0.000	0
3	4	1	0.000	0	3	10	1	0.000	0
3	4	2	0.000	0	3	10	2	0.000	0
3	4	3	0.000	0	3	10	3	0.000	0
3	4	4	0.000	0	3	10	4	0.000	0
3	4	5	0.000	0	3	10	5	0.000	0
3	4	6	0.000	0	3	10	6	0.000	0
3	4	7	0.000	0	3	10	7	0.000	0

Data Set 4

State	Occ	Yr	Exp.	Loss	State	Occ	Yr	Exp.	Loss
3	11	1	0.000	0	3	17	1	14.126	0
3	11	2	0.000	0	3	17	2	13.781	0
3	11	3	0.000	0	3	17	3	12.794	1
3	11	4	0.000	0	3	17	4	25.192	0
3	11	5	0.000	0	3	17	5	10.066	1
3	11	6	0.000	0	3	17	6	7.010	0
3	11	7	0.000	0	3	17	7	0.000	0
3	12	1	179.691	9	3	18	1	0.000	0
3	12	2	149.014	23	3	18	2	4.041	0
3	12	3	149.039	23	3	18	3	3.680	0
3	12	4	161.229	21	3	18	4	2.667	4
3	12	5	132.191	21	3	18	5	0.889	0
3	12	6	150.457	36	3	18	6	2.629	1
3	12	7	46.253	8	3	18	7	2.386	0
3	13	1	29.845	1	3	19	1	0.000	0
3	13	2	23.560	8	3	19	2	0.000	0
3	13	3	19.338	3	3	19	3	0.000	0
3	13	4	18.698	3	3	19	4	0.000	0
3	13	5	9.212	2	3	19	5	0.000	0
3	13	6	16.157	5	3	19	6	0.000	0
3	13	7	14.189	4	3	19	7	0.000	0
3	14	1	2.093	0	3	20	1	0.000	0
3	14	2	2.870	0	3	20	2	0.000	0
3	14	3	3.406	0	3	20	3	0.000	0
3	14	4	0.221	0	3	20	4	0.000	0
3	14	5	0.201	0	3	20	5	0.000	0
3	14	6	0.179	0	3	20	6	0.000	0
3	14	7	0.000	0	3	20	7	0.000	0
3	15	1	0.673	0	3	21	1	0.000	0
3	15	2	0.477	0	3	21	2	0.000	0
3	15	3	1.450	0	3	21	3	0.000	0
3	15	4	0.810	0	3	21	4	0.000	0
3	15	5	0.829	0	3	21	5	0.000	0
3	15	6	0.000	0	3	21	6	0.000	0
3	15	7	0.000	0	3	21	7	0.000	0
3	16	1	0.000	0	3	22	1	0.197	0
3	16	2	0.000	0	3	22	2	0.347	0
3	16	3	0.000	0	3	22	3	0.181	0
3	16	4	0.000	0	3	22	4	0.281	0
3	16	5	0.000	0	3	22	5	0.043	0
3	16	6	0.000	0	3	22	6	0.425	0
3	16	7	0.000	0	3	22	7	0.274	0

Data Set 4

State	Occ	Yr	Exp.	Loss	State	Occ	Yr	Exp.	Loss
3	23	1	3.172	0	4	4	1	0.000	0
3	23	2	3.233	0	4	4	2	0.000	0
3	23	3	2.661	0	4	4	3	0.009	0
3	23	4	0.951	0	4	4	4	0.000	0
3	23	5	2.165	0	4	4	5	0.002	0
3	23	6	1.313	0	4	4	6	1.875	0
3	23	7	3.496	0	4	4	7	0.450	0
3	24	1	0.000	0	4	5	1	509.027	25
3	24	2	0.000	0	4	5	2	386.262	47
3	24	3	0.000	0	4	5	3	450.24	34
3	24	4	0.641	0	4	5	4	459.818	43
3	24	5	0.583	1	4	5	5	486.224	29
3	24	6	0.000	0	4	5	6	547.395	69
3	24	7	0.432	0	4	5	7	427.830	35
3	25	1	13.702	0	4	6	1	466.568	118
3	25	2	10.806	0	4	6	2	356.964	90
3	25	3	10.104	0	4	6	3	452.247	117
3	25	4	10.809	0	4	6	4	421.831	117
3	25	5	9.240	1	4	6	5	446.563	141
3	25	6	8.746	0	4	6	6	505.293	145
3	25	7	7.204	0	4	6	7	467.767	101
4	1	1	61.915	3	4	7	1	0.000	0
4	1	2	53.792	10	4	7	2	0.000	0
4	1	3	72.029	8	4	7	3	0.000	0
4	1	4	88.613	8	4	7	4	0.000	0
4	1	5	74.504	6	4	7	5	0.000	0
4	1	6	100.845	7	4	7	6	0.000	0
4	1	7	80.335	5	4	7	7	0.000	0
4	2	1	86.020	10	4	8	1	40.104	15
4	2	2	130.682	7	4	8	2	50.330	14
4	2	3	85.489	3	4	8	3	48.484	4
4	2	4	51.209	3	4	8	4	43.418	5
4	2	5	52.260	6	4	8	5	31.351	4
4	2	6	76.560	7	4	8	6	33.473	4
4	2	7	66.118	3	4	8	7	22.522	4
4	3	1	303.446	15	4	9	1	10.278	2
4	3	2	365.504	14	4	9	2	10.650	0
4	3	3	376.006	19	4	9	3	9.698	3
4	3	4	381.427	13	4	9	4	9.046	1
4	3	5	335.017	14	4	9	5	6.301	0
4	3	6	480.300	30	4	9	6	10.151	0
4	3	7	384.625	23	4	9	7	8.522	0

Data Set 4

State	Occ	Yr	Exp.	Loss	State	Occ	Yr	Exp.	Loss
4	10	1	18.748	1	4	16	1	169.264	5
4	10	2	14.292	1	4	16	2	155.906	7
4	10	3	14.273	3	4	16	3	152.671	8
4	10	4	31.598	1	4	16	4	148.361	18
4	10	5	15.966	2	4	16	5	175.832	13
4	10	6	41.549	1	4	16	6	237.751	9
4	10	7	17.860	0	4	16	7	176.980	17
4	11	1	207.632	13	4	17	1	454.799	24
4	11	2	184.278	16	4	17	2	514.791	25
4	11	3	175.719	17	4	17	3	524.390	30
4	11	4	187.231	11	4	17	4	555.289	41
4	11	5	182.898	19	4	17	5	534.206	41
4	11	6	143.591	16	4	17	6	393.451	34
4	11	7	161.568	11	4	17	7	364.542	22
4	12	1	1219.753	60	4	18	1	16.338	2
4	12	2	1209.114	65	4	18	2	8.940	0
4	12	3	1319.611	92	4	18	3	11.172	0
4	12	4	1218.995	83	4	18	4	3.576	0
4	12	5	1147.229	46	4	18	5	11.425	0
4	12	6	1215.796	103	4	18	6	51.336	0
4	12	7	871.595	54	4	18	7	213.580	0
4	13	1	346.288	28	4	19	1	29.598	0
4	13	2	326.333	24	4	19	2	34.949	2
4	13	3	366.559	13	4	19	3	63.727	2
4	13	4	386.619	25	4	19	4	87.988	8
4	13	5	449.229	26	4	19	5	147.580	8
4	13	6	544.361	25	4	19	6	166.381	11
4	13	7	382.083	13	4	19	7	255.725	9
4	14	1	388.084	16	4	20	1	3724.839	211
4	14	2	422.299	25	4	20	2	3074.609	260
4	14	3	434.016	20	4	20	3	3521.765	183
4	14	4	436.244	26	4	20	4	3730.029	191
4	14	5	448.991	29	4	20	5	2954.277	164
4	14	6	605.180	23	4	20	6	2925.587	166
4	14	7	429.197	20	4	20	7	2231.543	117
4	15	1	411.307	35	4	21	1	1038.177	41
4	15	2	461.597	28	4	21	2	1036.252	37
4	15	3	513.164	45	4	21	3	1016.157	51
4	15	4	530.463	44	4	21	4	971.539	40
4	15	5	504.744	42	4	21	5	1187.021	47
4	15	6	378.033	39	4	21	6	1247.860	57
4	15	7	318.662	21	4	21	7	1120.058	45

Data Set 4

State	Occ	Yr	Exp.	Loss	State	Occ	Yr	Exp.	Loss
4	22	1	2549.965	164	5	3	1	358.581	11
4	22	2	2199.854	150	5	3	2	368.806	13
4	22	3	2348.835	149	5	3	3	383.699	10
4	22	4	2631.886	165	5	3	4	409.500	17
4	22	5	2434.569	124	5	3	5	454.812	15
4	22	6	3061.616	206	5	3	6	423.885	16
4	22	7	1732.496	98	5	3	7	447.185	12
4	23	1	299.075	5	5	4	1	0.000	0
4	23	2	322.514	3	5	4	2	0.062	0
4	23	3	295.618	8	5	4	3	0.343	0
4	23	4	333.570	20	5	4	4	0.205	0
4	23	5	367.030	18	5	4	5	0.119	0
4	23	6	538.020	18	5	4	6	1.078	0
4	23	7	377.634	13	5	4	7	2.269	0
4	24	1	558.267	50	5	5	1	450.786	17
4	24	2	639.542	34	5	5	2	409.726	9
4	24	3	698.642	56	5	5	3	466.055	15
4	24	4	749.089	51	5	5	4	452.077	16
4	24	5	802.909	65	5	5	5	467.771	17
4	24	6	772.890	48	5	5	6	473.632	18
4	24	7	554.838	27	5	5	7	404.318	14
4	25	1	4597.259	76	5	6	1	31.975	4
4	25	2	4223.573	91	5	6	2	27.048	6
4	25	3	4551.019	112	5	6	3	28.741	7
4	25	4	4767.551	101	5	6	4	30.963	2
4	25	5	5017.009	153	5	6	5	37.094	3
4	25	6	5589.034	162	5	6	6	40.048	2
4	25	7	4460.135	127	5	6	7	34.236	9
5	1	1	65.401	9	5	7	1	0.000	0
5	1	2	78.700	10	5	7	2	0.000	0
5	1	3	82.118	5	5	7	3	0.000	0
5	1	4	81.525	3	5	7	4	0.000	0
5	1	5	92.068	8	5	7	5	0.000	0
5	1	6	103.913	10	5	7	6	0.000	0
5	1	7	104.721	11	5	7	7	0.000	0
5	2	1	2.509	0	5	8	1	1.608	0
5	2	2	44.063	3	5	8	2	1.386	0
5	2	3	63.051	2	5	8	3	1.527	0
5	2	4	59.340	6	5	8	4	11.003	0
5	2	5	76.350	2	5	8	5	12.183	0
5	2	6	69.476	0	5	8	6	32.390	3
5	2	7	77.664	4	5	8	7	33.147	1

Data Set 4

State	Occ	Yr	Exp.	Loss	State	Occ	Yr	Exp.	Loss
5	9	1	26.170	2	5	15	1	493.662	18
5	9	2	25.625	4	5	15	2	492.383	36
5	9	3	12.705	1	5	15	3	404.343	9
5	9	4	23.593	1	5	15	4	434.358	18
5	9	5	12.394	1	5	15	5	464.566	19
5	9	6	12.831	0	5	15	6	530.355	20
5	9	7	1.254	0	5	15	7	532.976	14
5	10	1	336.930	25	5	16	1	162.791	20
5	10	2	419.357	28	5	16	2	173.838	26
5	10	3	485.827	41	5	16	3	174.374	17
5	10	4	513.817	17	5	16	4	164.275	16
5	10	5	623.062	9	5	16	5	161.153	9
5	10	6	608.477	6	5	16	6	167.969	8
5	10	7	194.290	11	5	16	7	153.274	6
5	11	1	504.635	19	5	17	1	325.778	6
5	11	2	487.475	26	5	17	2	308.522	12
5	11	3	504.561	11	5	17	3	322.762	13
5	11	4	500.661	21	5	17	4	347.113	12
5	11	5	552.084	20	5	17	5	428.528	13
5	11	6	513.896	13	5	17	6	445.512	27
5	11	7	403.989	11	5	17	7	308.972	11
5	12	1	1359.690	44	5	18	1	287.238	6
5	12	2	1342.849	65	5	18	2	165.624	3
5	12	3	1406.562	54	5	18	3	247.293	3
5	12	4	1336.310	44	5	18	4	264.460	10
5	12	5	1285.126	27	5	18	5	358.468	15
5	12	6	1269.444	33	5	18	6	265.788	16
5	12	7	999.323	16	5	18	7	252.510	3
5	13	1	732.625	23	5	19	1	12.147	0
5	13	2	724.743	41	5	19	2	10.683	0
5	13	3	1003.513	31	5	19	3	9.953	0
5	13	4	934.696	31	5	19	4	9.928	0
5	13	5	955.646	41	5	19	5	9.490	0
5	13	6	907.229	35	5	19	6	7.584	0
5	13	7	762.619	26	5	19	7	7.452	0
5	14	1	258.286	9	5	20	1	0.262	0
5	14	2	245.009	16	5	20	2	0.719	0
5	14	3	269.443	16	5	20	3	28.448	2
5	14	4	239.171	2	5	20	4	33.132	1
5	14	5	263.785	5	5	20	5	34.673	0
5	14	6	272.670	5	5	20	6	34.046	0
5	14	7	313.815	7	5	20	7	29.857	0

Data Set 4

State	Occ	Yr	Exp.	Loss	State	Occ	Yr	Exp.	Loss
5	21	1	14.845	0	6	2	1	2.407	0
5	21	2	20.443	0	6	2	2	14.357	0
5	21	3	33.885	1	6	2	3	17.088	1
5	21	4	20.008	0	6	2	4	17.009	5
5	21	5	19.169	0	6	2	5	15.823	5
5	21	6	21.065	0	6	2	6	30.721	3
5	21	7	21.531	0	6	2	7	16.272	2
5	22	1	131.714	6	6	3	1	64.450	1
5	22	2	114.771	3	6	3	2	63.689	0
5	22	3	113.217	8	6	3	3	67.320	0
5	22	4	120.410	4	6	3	4	64.222	7
5	22	5	113.653	1	6	3	5	59.745	7
5	22	6	86.358	6	6	3	6	110.922	4
5	22	7	103.296	3	6	3	7	60.538	4
5	23	1	5.400	1	6	4	1	0.000	0
5	23	2	4.897	0	6	4	2	1.210	0
5	23	3	10.134	0	6	4	3	2.114	0
5	23	4	11.436	1	6	4	4	2.538	0
5	23	5	16.488	2	6	4	5	2.361	0
5	23	6	17.596	3	6	4	6	5.584	1
5	23	7	19.464	1	6	4	7	3.871	0
5	24	1	7.302	0	6	5	1	341.309	26
5	24	2	8.013	0	6	5	2	290.222	19
5	24	3	4.095	0	6	5	3	257.047	23
5	24	4	10.817	1	6	5	4	182.091	23
5	24	5	12.136	0	6	5	5	169.397	23
5	24	6	10.864	0	6	5	6	333.409	33
5	24	7	2.652	0	6	5	7	152.281	7
5	25	1	750.649	5	6	6	1	463.010	75
5	25	2	846.588	11	6	6	2	416.012	66
5	25	3	851.739	13	6	6	3	430.732	85
5	25	4	825.799	6	6	6	4	425.154	93
5	25	5	774.472	6	6	6	5	395.516	93
5	25	6	682.399	12	6	6	6	556.077	164
5	25	7	618.494	9	6	6	7	272.694	56
6	1	1	10.272	8	6	7	1	0.000	0
6	1	2	31.655	9	6	7	2	0.000	0
6	1	3	36.823	4	6	7	3	0.000	0
6	1	4	35.682	2	6	7	4	0.000	0
6	1	5	33.194	2	6	7	5	0.000	0
6	1	6	58.022	6	6	7	6	0.000	0
6	1	7	30.921	9	6	7	7	0.000	0

Data Set 4

State	Occ	Yr	Exp.	Loss	State	Occ	Yr	Exp.	Loss
6	8	1	27.048	6	6	14	1	21.500	0
6	8	2	24.558	0	6	14	2	15.746	0
6	8	3	24.758	0	6	14	3	25.824	5
6	8	4	15.717	0	6	14	4	13.318	1
6	8	5	14.621	0	6	14	5	12.389	1
6	8	6	27.207	6	6	14	6	23.083	3
6	8	7	8.844	1	6	14	7	9.305	0
6	9	1	0.399	0	6	15	1	60.228	2
6	9	2	0.205	1	6	15	2	65.443	3
6	9	3	9.977	1	6	15	3	70.654	10
6	9	4	0.209	0	6	15	4	68.166	7
6	9	5	0.195	0	6	15	5	63.414	7
6	9	6	0.660	1	6	15	6	103.152	24
6	9	7	0.233	0	6	15	7	55.601	8
6	10	1	0.000	0	6	16	1	235.789	19
6	10	2	0.059	0	6	16	2	208.928	11
6	10	3	0.075	0	6	16	3	254.517	20
6	10	4	0.762	0	6	16	4	244.618	20
6	10	5	0.709	0	6	16	5	227.566	20
6	10	6	0.343	1	6	16	6	479.115	30
6	10	7	0.014	0	6	16	7	237.351	12
6	11	1	47.822	1	6	17	1	54.225	0
6	11	2	50.994	2	6	17	2	57.733	2
6	11	3	53.368	2	6	17	3	56.470	6
6	11	4	52.696	6	6	17	4	63.958	10
6	11	5	49.022	6	6	17	5	59.499	10
6	11	6	83.196	7	6	17	6	104.994	16
6	11	7	39.216	5	6	17	7	74.310	11
6	12	1	304.079	13	6	18	1	1.024	0
6	12	2	310.925	11	6	18	2	1.096	0
6	12	3	320.788	14	6	18	3	1.181	0
6	12	4	345.307	17	6	18	4	1.570	0
6	12	5	321.235	17	6	18	5	1.460	0
6	12	6	407.752	27	6	18	6	3.019	0
6	12	7	289.260	16	6	18	7	1.680	0
6	13	1	62.253	2	6	19	1	0.000	0
6	13	2	65.063	7	6	19	2	0.000	0
6	13	3	110.066	5	6	19	3	0.000	0
6	13	4	75.693	5	6	19	4	0.000	0
6	13	5	70.416	5	6	19	5	0.000	0
6	13	6	67.914	4	6	19	6	0.000	0
6	13	7	36.219	0	6	19	7	0.000	0

Data Set 4

State	Occ	Yr	Exp.	Loss	State	Occ	Yr	Exp.	Loss
6	20	1	152.344	14	7	1	1	19.396	3
6	20	2	110.593	1	7	1	2	20.417	2
6	20	3	165.334	8	7	1	3	19.960	2
6	20	4	206.055	14	7	1	4	18.658	0
6	20	5	191.690	14	7	1	5	16.459	4
6	20	6	326.149	36	7	1	6	15.234	0
6	20	7	190.739	12	7	1	7	13.363	3
6	21	1	91.209	2	7	2	1	21.636	1
6	21	2	103.779	7	7	2	2	31.749	0
6	21	3	132.512	9	7	2	3	94.324	6
6	21	4	126.576	6	7	2	4	83.996	3
6	21	5	117.752	6	7	2	5	99.267	4
6	21	6	185.731	6	7	2	6	85.056	3
6	21	7	112.738	1	7	2	7	89.722	12
6	22	1	312.522	18	7	3	1	80.046	2
6	22	2	343.011	20	7	3	2	87.639	3
6	22	3	442.907	18	7	3	3	89.863	1
6	22	4	376.127	25	7	3	4	114.644	5
6	22	5	349.907	25	7	3	5	114.938	4
6	22	6	789.973	73	7	3	6	99.181	3
6	22	7	337.793	9	7	3	7	99.830	6
6	23	1	6.520	0	7	4	1	0.000	0
6	23	2	3.983	0	7	4	2	0.000	0
6	23	3	2.107	0	7	4	3	0.000	0
6	23	4	2.872	0	7	4	4	0.000	0
6	23	5	2.671	0	7	4	5	1.291	0
6	23	6	2.670	0	7	4	6	0.000	0
6	23	7	6.369	0	7	4	7	0.000	0
6	24	1	0.512	0	7	5	1	178.266	8
6	24	2	0.000	0	7	5	2	168.032	11
6	24	3	0.000	0	7	5	3	137.926	10
6	24	4	0.000	0	7	5	4	136.791	4
6	24	5	0.000	0	7	5	5	156.900	6
6	24	6	0.000	0	7	5	6	142.182	11
6	24	7	0.037	0	7	5	7	146.341	3
6	25	1	379.584	12	7	6	1	1.765	0
6	25	2	297.539	7	7	6	2	2.067	0
6	25	3	268.807	6	7	6	3	0.918	0
6	25	4	364.761	9	7	6	4	1.056	0
6	25	5	339.333	9	7	6	5	1.704	0
6	25	6	583.105	24	7	6	6	1.728	0
6	25	7	367.863	17	7	6	7	2.218	1

Data Set 4

State	Occ	Yr	Exp.	Loss	State	Occ	Yr	Exp.	Loss
7	7	1	0.000	0	7	13	1	125.982	2
7	7	2	0.000	0	7	13	2	128.996	5
7	7	3	0.000	0	7	13	3	131.727	1
7	7	4	0.000	0	7	13	4	141.328	8
7	7	5	0.000	0	7	13	5	147.292	7
7	7	6	0.000	0	7	13	6	144.881	8
7	7	7	0.000	0	7	13	7	108.904	6
7	8	1	0.000	0	7	14	1	350.718	18
7	8	2	0.000	0	7	14	2	322.958	18
7	8	3	0.000	0	7	14	3	291.724	10
7	8	4	0.000	0	7	14	4	320.518	13
7	8	5	0.000	0	7	14	5	338.029	21
7	8	6	0.000	0	7	14	6	379.172	15
7	8	7	0.000	0	7	14	7	311.217	9
7	9	1	2.827	0	7	15	1	896.706	35
7	9	2	4.881	0	7	15	2	1033.142	58
7	9	3	5.038	0	7	15	3	1097.595	48
7	9	4	0.849	0	7	15	4	1253.944	67
7	9	5	2.459	0	7	15	5	1190.675	85
7	9	6	1.012	0	7	15	6	1240.498	92
7	9	7	0.000	0	7	15	7	1399.892	60
7	10	1	11.173	1	7	16	1	11.612	0
7	10	2	9.209	0	7	16	2	10.627	3
7	10	3	10.918	0	7	16	3	2.848	0
7	10	4	14.006	2	7	16	4	2.503	0
7	10	5	14.093	0	7	16	5	13.748	0
7	10	6	13.176	0	7	16	6	14.556	0
7	10	7	15.205	1	7	16	7	13.972	0
7	11	1	133.650	5	7	17	1	88.624	1
7	11	2	133.646	9	7	17	2	99.552	9
7	11	3	122.905	8	7	17	3	97.941	6
7	11	4	125.637	4	7	17	4	103.589	6
7	11	5	121.756	4	7	17	5	114.262	1
7	11	6	132.672	10	7	17	6	110.523	8
7	11	7	102.431	4	7	17	7	114.854	7
7	12	1	511.129	13	7	18	1	32.537	6
7	12	2	520.963	32	7	18	2	32.784	1
7	12	3	550.904	16	7	18	3	36.180	2
7	12	4	563.641	37	7	18	4	0.000	0
7	12	5	529.147	16	7	18	5	0.000	0
7	12	6	571.476	21	7	18	6	0.000	0
7	12	7	379.403	16	7	18	7	0.000	0

Data Set 4

State	Occ	Yr	Exp.	Loss	State	Occ	Yr	Exp.	Loss
7	19	1	0.000	0	7	25	1	113.914	0
7	19	2	0.000	0	7	25	2	114.830	4
7	19	3	0.000	0	7	25	3	154.929	6
7	19	4	0.000	0	7	25	4	152.695	2
7	19	5	0.000	0	7	25	5	140.777	0
7	19	6	0.000	0	7	25	6	162.506	1
7	19	7	0.000	0	7	25	7	129.075	1
7	20	1	0.000	0	8	1	1	17.029	7
7	20	2	0.000	0	8	1	2	201.942	35
7	20	3	1.391	0	8	1	3	244.090	51
7	20	4	1.089	0	8	1	4	226.271	47
7	20	5	0.674	0	8	1	5	121.338	15
7	20	6	1.048	0	8	1	6	125.167	19
7	20	7	1.202	0	8	1	7	134.714	14
7	21	1	5.473	0	8	2	1	7.241	0
7	21	2	6.440	0	8	2	2	223.756	81
7	21	3	1.051	0	8	2	3	227.306	56
7	21	4	5.138	2	8	2	4	227.627	52
7	21	5	5.690	0	8	2	5	103.045	16
7	21	6	5.715	0	8	2	6	161.215	60
7	21	7	6.934	0	8	2	7	172.778	40
7	22	1	17.601	1	8	3	1	195.436	34
7	22	2	16.108	4	8	3	2	242.608	51
7	22	3	18.352	0	8	3	3	264.067	25
7	22	4	17.392	2	8	3	4	306.318	37
7	22	5	15.083	1	8	3	5	159.131	17
7	22	6	8.891	0	8	3	6	283.996	38
7	22	7	5.378	0	8	3	7	345.801	34
7	23	1	0.000	0	8	4	1	104.460	7
7	23	2	1.209	0	8	4	2	132.383	13
7	23	3	1.284	0	8	4	3	124.146	8
7	23	4	1.289	0	8	4	4	141.509	8
7	23	5	0.968	0	8	4	5	61.220	3
7	23	6	1.707	0	8	4	6	68.865	4
7	23	7	2.816	0	8	4	7	79.169	8
7	24	1	0.817	0	8	5	1	984.752	219
7	24	2	0.481	0	8	5	2	659.618	130
7	24	3	1.052	0	8	5	3	616.913	108
7	24	4	0.691	0	8	5	4	645.042	98
7	24	5	0.839	0	8	5	5	279.972	56
7	24	6	1.907	0	8	5	6	322.667	80
7	24	7	0.888	0	8	5	7	454.643	54

Data Set 4

State	Occ	Yr	Exp.	Loss	State	Occ	Yr	Exp.	Loss
8	6	1	1962.686	545	8	12	1	663.110	57
8	6	2	2073.578	545	8	12	2	728.924	63
8	6	3	1999.936	549	8	12	3	593.602	56
8	6	4	2008.325	551	8	12	4	623.466	49
8	6	5	1082.922	292	8	12	5	342.459	21
8	6	6	1770.858	470	8	12	6	459.594	57
8	6	7	1700.566	400	8	12	7	630.510	59
8	7	1	0.000	0	8	13	1	162.172	14
8	7	2	0.000	0	8	13	2	157.167	27
8	7	3	0.000	0	8	13	3	150.148	22
8	7	4	0.000	0	8	13	4	65.356	3
8	7	5	0.000	0	8	13	5	79.811	10
8	7	6	0.000	0	8	13	6	105.833	22
8	7	7	0.000	0	8	13	7	127.045	8
8	8	1	4.689	2	8	14	1	106.084	14
8	8	2	4.584	5	8	14	2	107.927	21
8	8	3	5.105	3	8	14	3	84.418	9
8	8	4	1.632	0	8	14	4	94.295	15
8	8	5	0.925	0	8	14	5	31.739	7
8	8	6	1.735	0	8	14	6	86.841	14
8	8	7	0.769	0	8	14	7	107.188	11
8	9	1	0.541	0	8	15	1	222.246	49
8	9	2	0.146	0	8	15	2	241.576	65
8	9	3	2.799	0	8	15	3	189.392	38
8	9	4	1.835	1	8	15	4	130.845	25
8	9	5	0.364	0	8	15	5	84.708	19
8	9	6	3.689	1	8	15	6	122.229	39
8	9	7	14.258	5	8	15	7	125.142	18
8	10	1	21.290	3	8	16	1	542.872	115
8	10	2	24.922	6	8	16	2	524.168	88
8	10	3	29.303	4	8	16	3	583.582	121
8	10	4	30.794	2	8	16	4	480.739	182
8	10	5	32.255	3	8	16	5	346.022	155
8	10	6	44.079	2	8	16	6	633.092	159
8	10	7	43.120	6	8	16	7	565.911	130
8	11	1	174.921	20	8	17	1	315.500	27
8	11	2	197.949	26	8	17	2	330.438	37
8	11	3	211.725	41	8	17	3	306.497	46
8	11	4	260.470	36	8	17	4	316.470	50
8	11	5	206.783	29	8	17	5	126.058	23
8	11	6	292.633	34	8	17	6	217.512	48
8	11	7	258.186	24	8	17	7	278.349	41

Data Set 4

State	Occ	Yr	Exp.	Loss	State	Occ	Yr	Exp.	Loss
8	18	1	44.146	3	8	24	1	1.944	0
8	18	2	44.152	8	8	24	2	1.421	0
8	18	3	43.980	2	8	24	3	2.096	0
8	18	4	43.871	5	8	24	4	2.571	0
8	18	5	32.837	0	8	24	5	1.663	0
8	18	6	54.179	3	8	24	6	1.536	1
8	18	7	65.771	11	8	24	7	1.596	0
8	19	1	0.000	0	8	25	1	268.956	27
8	19	2	0.000	0	8	25	2	297.513	19
8	19	3	0.000	0	8	25	3	287.628	21
8	19	4	0.000	0	8	25	4	150.571	21
8	19	5	0.000	0	8	25	5	83.892	12
8	19	6	0.000	0	8	25	6	125.600	16
8	19	7	0.000	0	8	25	7	193.976	18
8	20	1	0.275	0	9	1	1	8.640	1
8	20	2	0.221	0	9	1	2	9.454	0
8	20	3	0.000	0	9	1	3	13.526	1
8	20	4	0.000	0	9	1	4	11.526	2
8	20	5	0.000	0	9	1	5	12.125	0
8	20	6	0.000	0	9	1	6	14.577	0
8	20	7	0.000	0	9	1	7	15.228	1
8	21	1	17.755	1	9	2	1	1.381	0
8	21	2	19.463	1	9	2	2	12.411	0
8	21	3	18.395	0	9	2	3	24.455	6
8	21	4	13.237	1	9	2	4	47.556	2
8	21	5	10.905	1	9	2	5	64.733	1
8	21	6	10.339	0	9	2	6	67.922	1
8	21	7	11.000	0	9	2	7	67.580	3
8	22	1	53.155	11	9	3	1	84.293	7
8	22	2	58.514	10	9	3	2	78.969	2
8	22	3	59.862	10	9	3	3	94.277	0
8	22	4	20.984	2	9	3	4	97.584	2
8	22	5	5.260	0	9	3	5	112.845	3
8	22	6	18.544	3	9	3	6	119.928	0
8	22	7	20.506	5	9	3	7	114.158	0
8	23	1	1.024	0	9	4	1	0.000	0
8	23	2	1.712	0	9	4	2	0.000	0
8	23	3	2.326	0	9	4	3	0.000	0
8	23	4	1.935	1	9	4	4	0.000	0
8	23	5	2.076	0	9	4	5	0.000	0
8	23	6	3.843	0	9	4	6	0.000	0
8	23	7	2.528	0	9	4	7	0.000	0

Data Set 4

State	Occ	Yr	Exp.	Loss	State	Occ	Yr	Exp.	Loss
9	5	1	103.141	15	9	11	1	61.887	3
9	5	2	133.114	3	9	11	2	63.222	3
9	5	3	98.666	3	9	11	3	71.104	3
9	5	4	61.518	3	9	11	4	86.664	5
9	5	5	32.922	0	9	11	5	77.103	4
9	5	6	32.661	0	9	11	6	98.791	3
9	5	7	33.668	0	9	11	7	113.671	0
9	6	1	20.513	3	9	12	1	336.036	17
9	6	2	11.868	3	9	12	2	328.650	12
9	6	3	8.889	2	9	12	3	352.083	13
9	6	4	7.733	0	9	12	4	352.578	14
9	6	5	8.104	5	9	12	5	340.921	9
9	6	6	7.803	0	9	12	6	358.731	8
9	6	7	7.819	0	9	12	7	369.883	15
9	7	1	0.000	0	9	13	1	84.858	3
9	7	2	0.000	0	9	13	2	84.104	3
9	7	3	0.000	0	9	13	3	91.425	2
9	7	4	0.000	0	9	13	4	92.225	3
9	7	5	0.000	0	9	13	5	95.470	13
9	7	6	0.000	0	9	13	6	95.686	5
9	7	7	0.000	0	9	13	7	104.998	3
9	8	1	0.000	0	9	14	1	96.505	5
9	8	2	0.000	0	9	14	2	99.129	4
9	8	3	0.000	0	9	14	3	111.033	6
9	8	4	0.000	0	9	14	4	110.513	2
9	8	5	0.000	0	9	14	5	101.611	7
9	8	6	0.000	0	9	14	6	103.022	2
9	8	7	0.000	0	9	14	7	101.516	3
9	9	1	158.887	6	9	15	1	67.683	3
9	9	2	216.146	8	9	15	2	76.516	1
9	9	3	270.930	19	9	15	3	77.213	7
9	9	4	349.445	21	9	15	4	80.675	3
9	9	5	330.062	21	9	15	5	86.761	3
9	9	6	333.017	19	9	15	6	72.906	1
9	9	7	392.271	27	9	15	7	72.961	2
9	10	1	68.791	2	9	16	1	4.709	0
9	10	2	69.051	4	9	16	2	3.457	0
9	10	3	79.363	2	9	16	3	8.081	1
9	10	4	82.433	1	9	16	4	5.886	0
9	10	5	117.813	1	9	16	5	4.730	0
9	10	6	107.257	3	9	16	6	6.774	0
9	10	7	87.217	2	9	16	7	6.258	0

Data Set 4

State	Occ	Yr	Exp.	Loss	State	Occ	Yr	Exp.	Loss
9	17	1	71.729	2	9	23	1	1.609	0
9	17	2	81.266	1	9	23	2	0.257	0
9	17	3	93.007	2	9	23	3	0.532	0
9	17	4	96.124	0	9	23	4	0.476	0
9	17	5	97.525	3	9	23	5	1.386	0
9	17	6	107.820	1	9	23	6	2.655	0
9	17	7	111.710	0	9	23	7	3.212	0
9	18	1	0.000	0	9	24	1	3.611	0
9	18	2	0.000	0	9	24	2	3.851	0
9	18	3	0.000	0	9	24	3	5.003	0
9	18	4	0.000	0	9	24	4	7.096	0
9	18	5	0.000	0	9	24	5	6.795	0
9	18	6	0.000	0	9	24	6	9.185	0
9	18	7	0.859	0	9	24	7	5.081	0
9	19	1	0.000	0	9	25	1	403.071	4
9	19	2	0.000	0	9	25	2	406.645	6
9	19	3	0.000	0	9	25	3	448.011	4
9	19	4	0.000	0	9	25	4	436.279	8
9	19	5	0.000	0	9	25	5	463.186	3
9	19	6	0.000	0	9	25	6	417.369	5
9	19	7	0.000	0	9	25	7	372.119	2
9	20	1	0.000	0	10	1	1	5.230	0
9	20	2	0.000	0	10	1	2	10.076	0
9	20	3	0.000	0	10	1	3	0.000	0
9	20	4	0.000	0	10	1	4	0.000	0
9	20	5	0.020	0	10	1	5	0.065	0
9	20	6	0.000	0	10	1	6	0.000	0
9	20	7	0.000	0	10	1	7	0.340	0
9	21	1	1.580	0	10	2	1	10.916	0
9	21	2	3.550	0	10	2	2	4.877	0
9	21	3	3.282	0	10	2	3	4.909	0
9	21	4	3.120	0	10	2	4	5.054	0
9	21	5	3.523	0	10	2	5	6.697	1
9	21	6	2.335	0	10	2	6	4.417	0
9	21	7	1.074	0	10	2	7	3.404	0
9	22	1	25.934	1	10	3	1	9.915	0
9	22	2	19.470	3	10	3	2	13.293	1
9	22	3	21.647	0	10	3	3	17.269	0
9	22	4	24.480	0	10	3	4	18.344	3
9	22	5	21.505	1	10	3	5	17.758	1
9	22	6	21.178	0	10	3	6	32.623	4
9	22	7	24.792	0	10	3	7	24.629	2

Data Set 4

State	Occ	Yr	Exp.	Loss	State	Occ	Yr	Exp.	Loss
10	4	1	0.000	0	10	10	1	243.058	12
10	4	2	0.000	0	10	10	2	296.975	8
10	4	3	0.000	0	10	10	3	379.053	22
10	4	4	0.000	0	10	10	4	346.209	18
10	4	5	0.000	0	10	10	5	693.615	36
10	4	6	0.956	0	10	10	6	720.382	35
10	4	7	0.000	0	10	10	7	571.079	10
10	5	1	7.644	1	10	11	1	29.869	1
10	5	2	7.866	1	10	11	2	45.871	1
10	5	3	1.656	0	10	11	3	58.846	1
10	5	4	12.613	0	10	11	4	42.950	5
10	5	5	8.205	1	10	11	5	32.716	5
10	5	6	3.300	0	10	11	6	47.219	3
10	5	7	4.343	0	10	11	7	63.042	7
10	6	1	219.547	41	10	12	1	86.356	5
10	6	2	137.478	34	10	12	2	81.535	8
10	6	3	183.977	71	10	12	3	89.442	3
10	6	4	154.898	64	10	12	4	78.428	9
10	6	5	141.344	56	10	12	5	37.078	8
10	6	6	149.891	44	10	12	6	71.283	10
10	6	7	216.714	42	10	12	7	51.975	4
10	7	1	0.000	0	10	13	1	0.701	0
10	7	2	0.000	0	10	13	2	0.593	0
10	7	3	0.000	0	10	13	3	0.000	0
10	7	4	0.000	0	10	13	4	0.000	0
10	7	5	0.000	0	10	13	5	0.000	0
10	7	6	0.000	0	10	13	6	1.278	0
10	7	7	0.000	0	10	13	7	0.904	0
10	8	1	0.000	0	10	14	1	8.013	0
10	8	2	0.000	0	10	14	2	2.449	1
10	8	3	0.000	0	10	14	3	3.431	0
10	8	4	0.000	0	10	14	4	4.636	0
10	8	5	0.000	0	10	14	5	4.895	0
10	8	6	0.000	0	10	14	6	4.915	0
10	8	7	0.000	0	10	14	7	4.821	3
10	9	1	0.123	0	10	15	1	7.394	0
10	9	2	0.123	0	10	15	2	5.503	2
10	9	3	0.646	0	10	15	3	3.811	0
10	9	4	0.823	0	10	15	4	4.645	2
10	9	5	5.663	0	10	15	5	4.764	2
10	9	6	6.514	0	10	15	6	3.431	0
10	9	7	6.946	0	10	15	7	4.011	0

Data Set 4

State	Occ	Yr	Exp.	Loss
10	16	1	478.908	51
10	16	2	478.109	68
10	16	3	551.303	89
10	16	4	183.392	9
10	16	5	63.749	9
10	16	6	134.775	4
10	16	7	98.758	2
10	17	1	9.523	0
10	17	2	1.741	0
10	17	3	0.235	0
10	17	4	0.286	0
10	17	5	0.000	0
10	17	6	0.000	0
10	17	7	0.178	0
10	18	1	0.000	0
10	18	2	0.000	0
10	18	3	7.725	1
10	18	4	6.289	0
10	18	5	5.185	0
10	18	6	0.000	0
10	18	7	0.000	0
10	19	1	0.000	0
10	19	2	0.000	0
10	19	3	0.000	0
10	19	4	0.000	0
10	19	5	0.000	0
10	19	6	0.000	0
10	19	7	0.000	0
10	20	1	0.000	0
10	20	2	0.000	0
10	20	3	0.000	0
10	20	4	0.000	0
10	20	5	0.000	0
10	20	6	0.000	0
10	20	7	0.000	0
10	21	1	2.733	0
10	21	2	1.091	0
10	21	3	1.071	0
10	21	4	2.305	0
10	21	5	1.899	0
10	21	6	1.195	1
10	21	7	0.406	0

State	Occ	Yr	Exp.	Loss
10	22	1	0.536	0
10	22	2	2.011	0
10	22	3	1.190	0
10	22	4	0.866	0
10	22	5	0.656	0
10	22	6	0.894	0
10	22	7	2.566	0
10	23	1	1.227	0
10	23	2	2.526	0
10	23	3	2.005	0
10	23	4	0.541	0
10	23	5	2.121	0
10	23	6	0.362	0
10	23	7	0.276	0
10	24	1	0.368	0
10	24	2	0.000	0
10	24	3	0.000	0
10	24	4	0.000	0
10	24	5	0.408	0
10	24	6	0.000	0
10	24	7	0.000	0
10	25	1	3.014	0
10	25	2	2.010	0
10	25	3	3.143	0
10	25	4	1.767	0
10	25	5	2.264	0
10	25	6	3.388	0
10	25	7	3.804	0

Bibliography

Abramowitz, M. and Stegun, I., eds. (1964), *Handbook of Mathematical Functions with Formulas, Graphs, and Mathematical Tables*, New York: John Wiley.

Batten, R. (1978), *Mortality Table Construction*, Englewood Ciffs, NJ: Prentice-Hall.

Berger, J. (1985), *Bayesian Inference in Statistical Analysis, 2nd ed.*, New York: Springer-Verlag.

Bowers, N., Gerber, H., Hickman, J., Jones, D. and Nesbitt, C. (1986), *Actuarial Mathematics*, Chicago: Society of Actuaries.

Box, G. (1980), "Sampling and Bayes' Inference in Scientific Modelling and Robustness," *Journal of the Royal Statistical Society, Series A,* **143**, 383–404.

Box, G. (1983), "An Apology for Ecumensim in Statistics," in *Proceedings of the conference on Scientific Inference, Data Analysis, and Robustness*, G. Box, T. Leonard, and C-F Wu, eds., New York: Academic Press.

Broffitt, J. (1984), "A Bayes Estimator for Ordered Parameters and Isotonic Bayesian Graduation," *Scandinavian Actuarial Journal*, 231–247.

Bühlmann, H. (1967), "Experience Rating and Credibility," *ASTIN Bulletin*, **4**, 199–207.

Bühlmann, H. and Straub, E. (1972), "Credibility for Loss Ratios," English translation in *ARCH*, **1972.2.**

Burden, R. and Faires, J. (1989), *Numerical Analsysis,* 4th ed., Boston: PWS-Kent.

Carlin, B. (1991), "Analyzing Nonlinear and Non-Gaussian Actuarial Time Series," *ARCH*, to appear.

Carlin, B. (1992), "A Simple Monte Carlo Approach to Bayesian Graduation," *Transactions of the Society of Actuaries*, paper submitted.

Dennis, J. Jr., and Schnabel, R. (1983), *Numerical Methods for Unconstrained Optimization and Nonlinear Equations*, Englewood Cliffs, NJ: Prentice-Hall.

van Dijk, H. and Kloek, T. (1980), "Further Experience in Bayesian Analysis Using Monte Carlo Integration," *Journal of Econometrics*, **14**, 307–328.

Efron, B. (1986), "Why Isn't Everyone a Bayesian?" *The American Statistician,* **40**, 1–11 (including comments and reply).

Ericson, W. (1969), "A Note on the Posterior Mean of a Population Mean," *Journal of the Royal Statistical Society, Series B*, **31**, 332–334.

Gelfand, A. and Smith, A. (1990), "Sampling-Based Approaches to Calculating Marginal Densities," *Journal of the American Statistical Association*, **85**, 398–409.

Gersch, W. and Kitagawa, G. (1988), "Smoothness Priors in Time Series," in *Bayesian Analysis of Time Series and Dynamic Models,* J. Small, ed., New York: Marcel Dekker.

Goel, P. (1982), "On Implications of Credible Means Being Exact Bayesian," *Scandinavian Actuarial Journal*, 41–46.

Goovaerts, M. and Hoogstad, W. (1987), *Credibility Theory*, Surveys of Actuarial Studies No. 4, Rotterdam: Nationale-Nederlanden.

Graybill, F. (1961), *An Introduction to Linear Statistical Models*, New York: McGraw-Hill.

Hachemeister, C. (1975), "Credibility for Regression Models with Applications to Trend," in *Crediblity: Theory and Applications*, P. Kahn, ed., New York: Academic Press.

Harville, D. (1977), "Maximum Likelihood Approaches to Variance Component Estimation and to Related Problems," *Journal of the American Statistical Association,* **72**, 320–338.

Herzog, T., *An Introduction to Bayesian Crediblity and Related Topics, Part 4 Study Note*, New York: Casualty Actuarial Society.

Hewitt, C. Jr. (1967), "Loss Ratio Distributions – A Model," *Proceedings of the Casualty Actuarial Society*, **54**, 70–88.

Hogg, R. and Klugman, S. (1984), *Loss Distributions*, New York: Wiley.

Hossack, I., Pollard, J. and Zehnwirth, B. (1983), *Introductory Statistics with Applications in General Insurance*, Cambridge, England: Cambridge University Press.

Jeffreys, H. (1961), *Theory of Probability, 3rd ed.*, London: Oxford University Press.

Jewell, W. (1974), "Credible Means are Exact Bayesian for Exponential Families," *ASTIN Bulletin*, **8**, 77–90.

deJong, P. and Zehnwirth, B. (1983), "Credibility Theory and the Kalman Filter," *Insurance: Mathematics and Economics,* **2**, 281–286.

Kloek, T. and van Dijk, H. (1978), "Bayesian Estimates of Equation System Parameters: An Application of Integration by Monte Carlo," *Econometrica*, **46**, 1–19.

Klugman, S. (1985), "Distributional Aspects and Evaluation of Some Variance Estimators in Credibility Models," *ARCH,* **1985.1**, 73–97.

Klugman, S. (1987), "Credibility for Classification Ratemaking via the Hierarchical Normal Linear Model," *Proceedings of the Casualty Actuarial Society*, **74**, 272–321.

Klugman, S. (1989), "Bayesian Modelling of Mortality Catastrophes," *Insurance: Mathematics and Economics*, **8**, 159–164.

Klugman, S. (1989), "Measuring Uncertainty in Increased Limits Factors – A Bayesian Approach," *Proceedings of the XXIst ASTIN Colloquium,* 199–210.

Klugman, S. (1990), "Credibility for Increased Limits," *Insurance: Mathematics and Economics*, **9**, 77–80.

Ledolter, J., Klugman, S. and Lee, C. (1991), "Credibility Models with Time-Varying Trend Components," *ASTIN Bulletin*, **21**, 73–91.

Lindley, D. (1982), "Scoring Rules and the Inevitability of Probability," *International Statistics Review*, **50**, 1–26.

Lindley, D. (1983), "Theory and Practice of Bayesian Statistics," *The Statistician,* **32**, 1–12.

Lindley, D. (1987), "The Probability Approach to the Treatment of Uncertainty in Artificial Intelligence and Expert Systems," *Statistical Science*, **2**, 17–24.

Lindley, D. and Smith, A. (1972), "Bayes Estimates for the Linear Model," *Journal of the Royal Statistical Society, Series B*, **34**, 1–41.

London, R. (1985), *Graduation: The Revision of Estimates,* Winsted, CT: ACTEX.

London, R. (1988), *Survival Models and Their Estimation, 2nd ed.,* Winsted, CT: ACTEX.

Meinhold, R. and Singpurwalla, N. (1983), "Understanding the Kalman Filter," *The American Statistician*, **37**, 123–127

Meyers, G. (1984), "Empirical Bayesian Credibility for Workers' Compensation Classification Ratemaking," *Proceedings of the Casualty Actuarial Society*, **71**, 96–121.

Meyers, G. (1985), "An Analysis of Experience Rating," *Proceedings of the Casualty Actuarial Society*, **72**, 278–317.

Meyers, G., and Schenker, N. (1983), "Parameter Uncertainty in the Collective Risk Model," *Proceedings of the Casualty Actuarial Society*, **70**, 111–143.

Miller, R. and Fortney, W. (1984), "Industry-wide Expense Standards Using Random Coefficient Regression," *Insurance: Mathematics and Economics*, **3**, 19–33.

Morris, C. (1983), "Parametric Empirical Bayes Inference: Theory and Applications," *Journal of the American Statisitical Association,* **78**, 47–55.

Mowbray, A. (1914), "How Extensive a Payroll Exposure is Necessary to Give a Dependable Pure Premium?" *Proceedings of the Casualty Actuarial Society*, **1**, 24–30.

Naylor, J. and Smith, A. (1982), "Applications of a Method for the Efficient Computation of Posterior Distributions," *Applied Statistics*, **31**, 214–225.

Nelder, J. and Mead, R. (1965), "A Simplex Method for Function Minimization," *Computer Journal,* **6**, 308–313.

Parmenter, M. (1988), *Theory of Interest and Life Contingencies, with Pension Applications: A Problem Solving Approach*, Winsted, CT: ACTEX.

Philbrick, S. (1981), "An Examination of Credibility Concepts," *Proceedings of the Casualty Actuarial Society*, **68**, 195–219.

Racine, A., Grieve, A., Fluhler, H., and Smith, A. (1986), "Bayesian Methods in Practice: Experiences in the Pharmaceutical Industry," *Applied Statistics*, **35**, 93–120.

Schwartz, G. (1978), "Estimating the Dimension of a Model," *Annals of Statistics*, **6**, 461–464.

Smith, A. and West, M. (1983), "Monitoring Renal Transplants: An Application of the Multiprocess Kalman Filter," *Biometrics*, **39**, 867–878.

Society of Actuaries (1985), "1975-80 Basic Tables with Appendix of Age-Last-Birthday Basic Tables," *Transactions, Society of Actuaries, 1982 Reports,* 55–81.

Tierney, L. and Kadane, J. (1986), "Accurate Approximations for Posterior Moments and Marginal Densities," *Journal of the American Statistical Association*," **81**, 82–90.

Venter, G. (1985), "Structured Credibility in Applications — Hierarchical, Multidimensional, and Multivariate Models," *ARCH*, **1985.2**, 267–308.

Venter, G. (1986), "Classical Partial Credibility with Application to Trend", *Proceedings of the Casualty Actuarial Society*, **73**, 27–51.

DeVylder, F. (1981), "Practical Credibility Theory with Emphasis on Optimal Parameter Estimation," *ASTIN Bulletin*, **12**, 115–131.

DeVylder, F. (1986), "General Regression in Multidimensional Credibility Theory," in *Premium Calculation in Insurance*, M. Goovaerts, J. Haezendonck, and F. DeVylder, eds., Dordrecht, Holland: Reidel.

Whitney, A. (1918), "The Theory of Experience Rating," *Proceedings of the Casualty Actuarial Society*, **4**, 274–292.

INDEX

A note from the author

Copies of the GAUSS programs and Data sets 2–4 are available from the author as ASCII files. There are two methods of obtaining them:

1. Send a blank, formatted (for DOS) disk (both 3.5" and 5.25" as well as regular or high density are acceptable) along with a prepaid return mailer to the author at the following address:

Professor Stuart Klugman
College of Business and Public Administration
Drake University
2507 University Avenue
Des Moines, IA 50311

2. Send a message to me via electronic mail informing me of your return e-mail address (and indicating that you are requesting these programs and data). I am on BITNET and can also send materials on INTERNET. My BITNET address is SK0911R@DRAKE.BITNET. The BITNET extension may not be necessary, depending upon your system. The third character is the number zero, not the letter "oh."

This offer is good as long I retain the capability of making such copies. The addresses, both postal and electronic, cannot be guaranteed to remain accurate.